Interstellar Travel and Multi-Generation Space Ships

An interplanetary vehicle, based on Project Orion, using pulsed fission propulsion.
Courtesy of NASA

Interstellar Travel and Multi-Generation Space Ships

Edited by

Yoji Kondo, Frederick Bruhweiler,
John Moore and Charles Sheffield

An Apogee Books Publication

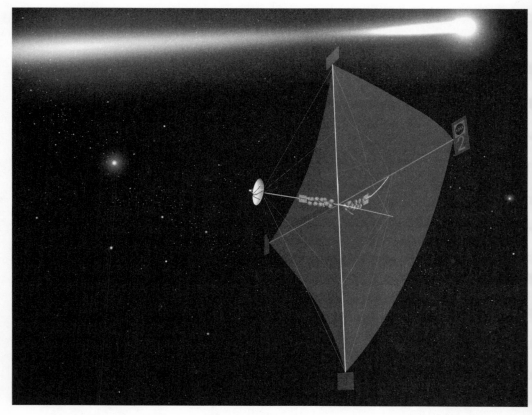

Lightsails technologies can utilize light, magnetic wind, or any of various beamed
energy sources to achieve acceleration without the need to carry onboard fuel.
Courtesy of NASA

All rights reserved under article two of the Berne Copyright Convention (1971).
We acknowledge the financial support of the Government of Canada through the Book Publishing Industry Development Program for our publishing activities.
Published by Apogee Books an imprint of Collector's Guide Publishing Inc., Box 62034, Burlington, Ontario, Canada, L7R 4K2, http://www.cgpublishing.com
Printed and bound in Canada

INTERSTELLAR TRAVEL AND MULTI-GENERATION SPACE SHIPS
Edited by
Y. Kondo, F.C. Bruhweiler,
J. Moore and C. Sheffield

ISBN 1-896522-99-8 — ISSN 1496-6921
Front cover illustration by Paul McGhee
©2003 Apogee Books

Contents

Dedication: To Charles Sheffield and Robert Forward 6
Preface
    by the Editors 6
Overview
    by Yoji Kondo 7

PART I: **Physical Sciences & Technology Section**

Fly Me to the Stars: Interstellar Travel in Fact and Fiction
    by Charles Sheffield 20
Ad Astra!
    by Robert Forward 29
The Ultimate Exploration: A Review of Propulsion Concepts for Interstellar Flight
    by Geoffrey Landis 52
Colonizing Other Words
    by Joe Haldeman 62
Why We Must Go
    by Doug Beason 69

PART II: **Anthropology, Genetics & Linguistics**

Kin-Based Crews for Interstellar Multi-Generation Space Travel
    by John H. Moore 80
Genetic Considerations in Multi-Generational Space Travel
    by Dennis H. O'Rourke 89
Language Change and Cultural Continuity on Multi-Generational Space Ships
    by Sarah G. Thomason 100

PART III: **Looking for Life in Unlikely Places**

Looking for Life in Unlikely Places
    by Freeman J. Dyson 105

PART IV: **Memories of Charles Sheffield and Robert Forward**

Remembering Charles Sheffield
    by Yoji Kondo 122
Reminiscences: Bob Forward
    by Geoffrey Landis 123

Dedication

This volume is dedicated to the memories of Charles Sheffield, who passed away on 2 November 2002, and Robert Forward, who died on 21 September 2002. Charles was 67 and Bob 70. Both had active careers in the space program and made important contributions. The current book contains their last articles on this topic. Charles also helped organize the AAAS symposium on *Interstellar Travel and Multi-Generation Space Ships*, held in Boston on 15 February 2002.

Preface

This book is based on the AAAS (American Association for the Advancement of Science) symposium *Interstellar Travel and Multi-Generation Space Ships* which was held in Boston on 15 February 2002.

There have been a number of publications on this subject. Most tended to focus on technical issues in designing interstellar space ships. In this book, we hope to have a meeting of the two fields that have often been considered separately. We shall discuss (I) scientific and engineering (what is often known as hard technical) issues related to interstellar travel, and (II) anthropological, genetic and linguistic (usually thought of as humanity-related) issues – together. We would be most pleased if the readers find that the twain have met on this fascinating subject.

— Yoji Kondo, Frederick Bruhweiler, John Moore, Charles Sheffield

Interstellar Travel and Multi-Generation Space Ships: An Overview

Yoji Kondo
NASA Goddard Space Flight Center
Greenbelt, MD 20771

In this opening chapter, I would like to present a brief overview of the subjects that are relevant to this AAAS (American Association for the Advancement of Science) symposium on *Interstellar Travel and Multi-Generation Space Ships*.

[I] Why Do We Want to Travel to Distant Planetary Systems?

Let us first consider the reasons why we want to talk about interstellar travel. Such an undertaking would likely dwarf all other undertakings in space, including the Apollo missions and much speculated manned missions to Mars. We do not yet have the resources or the technological means to realize interstellar travel, but there are at least two important reasons why it would make sense to start thinking about interstellar space ships now, despite the formidable nature of the ideas involved.

(A) Long-term survival of the human race

In order to survive as a race, we must get out of the solar system at some point in the future.

There are those among us who would be satisfied if their retirement years appeared reasonably secure. On the other hand, the majority of us would, I think, be concerned what kind of a society our children would live in. Some of us would even worry about the world our grandchildren would live in. The horizon for some would not seem to extend much further than that. For example, a scientist discussing the future of energy resources at a futurist conference, was happy to hear that the world petroleum resources, if shale oil is included, will not be depleted until the time of our great grandchildren; to him, there was no point in wasting our time talking about alternative sources of energy, such as solar power satellites. (I am not sure what he thought the children of our great grandchildren would be using for energy.)

Why don't we take a really long view of the human race to see what is in store for us on Earth?

As the Sun evolves to become a giant star, the environment of Earth will begin to change. The terrestrial environment will turn less clement for the survival of our species. It will be a long time before that change becomes really serious, but the sooner we confront the problems head on, the better we will be prepared when the inevitable arrives.

The Sun is a main sequence star; it is estimated to be about 4.5 billion years old. It presently provides a clement environment for human beings to live on Earth at a distance of some 150,000,000 kilometers (i.e., one astronomical unit) from it. It does so through conversion of hydrogen into helium at its core. However, the Sun is evolving and, over the next few billion years, will begin to become a giant star, as it exhausts the hydrogen fuel at its core and the internal balance begins to transform. After going through the giant star phase, the Sun will shed its outer atmospheres and reveal its core; it will become a white dwarf star, which is entirely unsuitable for maintaining human life on Earth or anywhere else in the solar system. The transformation of the Sun to a white dwarf star will take an additional several billion years, but, however long it takes, it will happen nevertheless.

The change will begin to affect Earth's environments negatively – and in time materially. Our mother planet will then be no longer hospitable for human habitation. It will become too hot; numerous giant solar flares from the Sun will be scorching our atmosphere.

We might terraform Mars and some of the Jovian or Saturnian moons and move farther away from the Sun's fury and survive a while. But even Edgeworth-Kuiper Belt objects, located beyond the most distant planet of the solar system, Pluto, would eventually become unsuitable for the survival of our species as the Sun turns into a white dwarf star after going through its giant phase.

At some point, then, it will be a good idea to move out of the solar system entirely. The time to start thinking about it is NOW. The reason for this imperative is that, once a culture turns inward, it is often difficult to force it to look outward again. Consider, as a historical example, the Ming Dynasty, which sent out an expeditionary fleet as far as the East Coast of Africa; this was done when the empire was young and full of energy at the beginning of the fifteenth century after ousting the Mongol overlords in the latter part of the fourteenth century. When this early outburst of energy and exuberance subsided, the Ming empire turned inward; no more overseas expeditions were sent thereafter. Ming was eventually conquered by the Tungus speaking Manchus in the mid-seventeenth century.

(B) Keeping our civilization vibrant by exploring the unknown

Exploring the unknown is essential for a civilization to remain alive. It is a cultural imperative, without which we, as a race, shall wither and decay. The humanoid creatures that remained in the caves and did not want to explore what might be beyond the mountain ranges, probably became extinct as the environment changed or were out-competed by those more adventurous. The spirit of exploring the unknown is especially deeply embedded in the American psyche. Not following this ethnic heritage could lead to a cultural suicide.

For the sake of maintaining our cultural virility, the time to start thinking seriously of interstellar travel is *now!* Unless we start making plans now, however incomplete they may be with our still incomplete scientific, technological, and anthropological knowledge, our civilization may turn inward, as did several civilizations in the past, and we may never start looking outward again. The question might be put succinctly as follows: If not now, when? Do we want to wait until our culture is in a state of constant decay through lack of cultural and physical stimulus to grow and explore?

Joe Haldeman considers in his article various reasons why human beings might want to migrate to other planetary systems.

II] Are There Other Earth-like Planets Elsewhere?

Do we know where else in the galaxy Earth-like planets are? The answer today is "No, we do not know – yet." Over the next few decades, however, the answer could very possibly change to "Yes!"

In recent years, over 60 planets have been detected orbiting around distant stars but most of them are Jupiter-sized massive planets. However, in 2007, Kepler observatory, which has been designed to detect Earth-like planets (both in size and in their orbital distances from their Sun-like primary stars), will be launched into an Earth-trailing orbit. In its four to five year mission, some 50 to 60 Earth-like planets are expected to be found orbiting around some 100,000 sun-like stars within the 105 square degrees in the Cygnus region of the sky. For more information about Kepler, visit the web site: www.kepler.arc.nasa.gov.

Earth-like stars to be detected in the Kepler Mission are hundreds of light years away from the Sun. However, over the next several decades, interferometric arrays of apodized telescopes in space, possibly located on the Moon, can start directly imaging Earth-like planets within tens of light years of the solar system. Long before we have star ships capable of journeying to distant solar systems, we can expect to know which way the ships need to be headed.

[III] Scientific and Technological Problems of Interstellar Travel

(A) Distances and travel time involved

Interstellar distances are immense compared to the distances within our own solar system. Expressed in terms of the speed of light, the distances from Earth to various objects in our solar system range from light seconds (Moon), to light minutes (Sun, Mercury, Venus, Mars, most asteroids, and Jupiter), and to light hours and days (Saturn, Uranus, Pluto, and Edgeworth-Kuiper Belt objects). In comparison, the distances to even the nearest stars are measured in light years, the nearest being Alpha Centauri which happens to be a binary system; the distance to it is 4.38 light years or 1.34 parsecs. One light year is just slightly below ten to the 18th (10^{18}) centimeters or ten to the 13th kilometers – ten trillion kilometers. A space ship traveling at a constant velocity of 100 km/s, which is several times the escape velocity from Earth, will spend 10^{11} seconds or some three millennia covering such a distance.

If this distance appears unimpressive to you, our Milky Way galaxy is about 100,000 light years across its lens-shaped disk. Even the nearest galaxies, the Large and Small Magellanic Clouds, are more than one and a half times that distance away from us.

Charles Sheffield has a list of various nearby stellar systems in his article. The nearest of them all is the Alpha Centauri binary system. Its primary is a G2 V (unevolved main-sequence) star, as close to our Sun in spectral type as they come. It would appear to be an ideal place to look for an Earth-like planet with an environment supportive of life – except that it has a companion star, whose spectral type is K2 V; they orbit each other in a period of about 80 years at a distance something like that of Uranus from the Sun. Because of the orbital perturbations caused by this companion star, a stable orbit for an Earth-like planet around the G2 V primary would be quite unlikely. (Too bad!)

(B) Constant velocity ships (slow boat to China) versus continuously accelerating ships

(For detailed discussions, see articles by Robert Forward and Geoffrey Landis in this book.)

(B-1) Constant Velocity Ships

To reach our closest neighbor, Alpha Centauri A-B, even though it is unlikely to have a planet suitable for human colonization, it will take more than ten thousand years to get there – if we travel at a constant velocity of 100 km/s. That would imply several hundred generations for the crew.

This is the reason why the concept of multi-generation space ships is an issue relevant to interstellar travel. And, that is why, as an important part of this symposium, anthropologists will discuss what the optimum-minimum number of crew composition would be for multi-generation space ships. To maintain a stable crew, we need not only to consider the gene pools, but also the cultural mix, as well as the social structure of the crew. Language for the crew would also be an important element – if only to avoid duplicating the problems encountered by the construction engineers and laborers of the Tower of Babel several millennia ago.

On the other hand, would it be possible to cover the interstellar distances without requiring multi-generation space ships? That would not necessarily be impossible, if we could accelerate space ships continuously. Even when we have continuously accelerating space ships, however, we will still need to understand the optimum-minimum number of population mix necessary for settling such distant planets. The anthropological considerations for multi-generation space ships discussed in the second part of this volume are still important and essential in making plans for interstellar migration of the human race.

(B-2) Continuous Acceleration Ships

We will first consider what will happen to the duration of the trip if we accelerate a ship continuously. We will then look at various engineering possibilities for achieving it.

If we accelerate at a rate of one gravitational force, i.e., about 980 centimeters per second per second, the ship can reach a speed close to that of the speed of light in about a year. The calculation goes as follows: the gravitational acceleration is a little shy of 10^3 (ten to the third) cm/s/s. The number of seconds in a year is a little over 3.15×10^7 seconds. If you multiply the two, you come up with a figure that is almost exactly the speed of light, i.e., 3×10^{10} cm/s – except that due to the relativistic effect on the time and distance involved the actual ship's velocity will never quite reach the speed of light when measured from the home base. However, the mathematical exercise above shows that in the course of a year – in just one year – the ship accelerated continuously at one g will approach the speed of light. The one g acceleration will of course be quite comfortable to the crew and it is immensely easier to keep the crew healthy in one-g environment than in free-fall or at an acceleration lower than one g.

When the ship approaches the speed of light, the relativistic time dilation will significantly shorten the time for (experienced by) the crew. So, if you accelerate the ship continuously to 99% of the speed of light, traveling to a planetary system some 100 years away will be a matter of only a few years to the crew, including the time necessary for acceleration and deceleration; the latter is necessary unless we want to zoom past the destination at almost the speed of light.

(B-3) Means for Continuous Acceleration

Space ships have limited cargo capacities, which makes carrying the fuel and reaction mass necessary for continuous acceleration very difficult indeed. The most compact form of energy we can conceive of today is anti-matter. When a quantity of matter encounters an equal quantity of anti-matter, the entire matter will be annihilated producing pure energy; the energy produced in a matter-antimatter conversion would be incomparably greater than that from a conversion of a very small fraction of the total matter achieved in the fusion of hydrogen atoms or in the splitting of uranium atoms.

Requirements to carry reaction mass on board could present another kind of serious problem for an interstellar ship. Newton's Third Law, describing the momentum transfer between the reaction mass ejected from the ship and the subsequent acceleration of the ship, would be strictly enforced by the Universe. One way to get around this problem is to use the pure energy from the matter-antimatter annihilation in the form of radiation pressure on the ship, obviating the necessity of any reaction mass. Photons from the matter-antimatter annihilation would be pushing the ship. One might call such a propulsion system a photon drive.

Alternatively, the energy could be transmitted to the ship in the form of beamed energy from the base where the energy is generated using a solar power station or possibly some form of nuclear power plant. That beamed energy could be used to accelerate reaction mass or in the form of a photon drive that would not require reaction mass to propel the ship.

(B-3a) Interstellar Ramjet and the Density of Interstellar Matter

Once the ship reaches a fair fraction of the speed of light, possibly using one of the propulsion technologies mentioned above, but not necessarily sufficient for the relativistic time dilation to become significant, the ship could scoop up interstellar matter as reaction mass and even possibly as fuel – the latter if we achieve controlled hydrogen fusion over the next few centuries. (We do not have that technology as yet, although it is the basic source of energy at the core of stars.) The scooping up may be performed using strong magnetic fields rather than an actual mechanical funnel; the latter method would likely add immensely to the total mass of the ship. Most of the matter in interstellar space is ionized, so a magnetic scoop might be an efficient way in (what Bob Bussard called) an interstellar ramjet. The density of the interstellar matter is relatively low in the solar vicinity, being about 0.1 particle per cubic centimeter, whereas the average for the galaxy is around 1 particle per cc. Even the latter density is a pretty good vacuum, but we could have an immense quantity for the scoop when the ship is traveling at a very high velocity. Actually, immediately beyond the solar neighborhood (with a radius in the range of a few tens of light years) is a region of more tenuous (density = 0.01 to 0.001 particle per cc) hot interstellar gas that extends possibly several tens of light years in some directions.

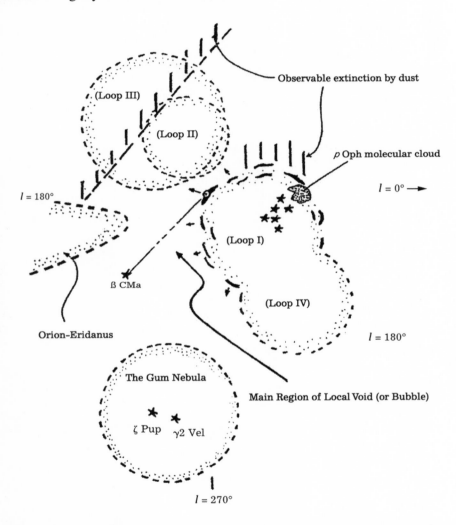

Figure 1 – The local interstellar medium within about 1,000 light years of the Sun.

The Sun is located near the center of the figure and is embedded in a small "island" of gas (shown as a small, dark patch connected by a line to the star Beta CMa) with a density of around 0.1 atom per cc. This cloud is on the edge of an old supernova remnant (Loop I). The gas interior of the Loops and other large-scale structures, including the main region of the Local Void, are filled by a pervasive, million-degree hot gas with a density in the range of 0.01 to 0.001 atom per cc. The Loops are thought to be old supernova remnants. Additional details, such as the very dense ρ (Greek letter "rho") Ophiuchi molecular cloud, are also shown. The large scale mean density of the interstellar matter within our galaxy is about 1 atom per cc. However, the typical element in the interstellar medium out to 1,000 light years is typified by densities of 0.01 to 0.001 atoms per cc. This means there must be very dense clouds in the galaxy such that the average density is roughly 1 atom per cc. [Bruhweiler, F.C., *Astrophysics in the Extreme Ultraviolet*, eds. S. Bowyer & R.F. Malina, Kluwer Academic Publishers, 1996, p.261]

The reader might also look up NASA CP-2345 (1984), *The Local Interstellar Medium*, International Astronomical Union Colloquium No. 81, edited by Kondo Bruhweiler and Savage, and several other papers on this subject that have been published in professional journals (such as the Astrophysical Journal or Astronomy and Astrophysics) since that meeting.

(B-3b) Collisions with meteoric and other solid matters in interstellar space

Another important issue that must be addressed in designing fast-traveling interstellar space ships is the possibility of their collisions with interstellar particles when traveling at high velocities. When the ship is traveling at 0.001c or one-thousandth the speed of light, i.e., 3×10^7 cm/s, a gram of pebble colliding with the ship head-on would produce an explosive energy of some 10^{15} erg, roughly equivalent to 20 kilograms of TNT.

Obviously, we need a protective measure against a catastrophe of that sort, however remote that possibility might be. Some form of protective shield placed in front of the ship could reduce the risk, although that might add considerably to the ship's dead-weight.

If the ship's design involves a vast magnetic scoop for the interstellar ramjet, the streams of ionized gases speeding toward the intake could carry with them small particles into the intake, which could be designed to digest and make use of gases as well as small particles.

(C) **Faster than light ships**

The propulsion systems discussed in Section (B) do not require new physical principles, such as a warp drive, and do not travel at velocities in excess of the speed of light.

However, there are speculations about space ships that can travel faster than the speed of light, which involve physical principles that are unknown to us. These rather speculative means of interstellar transportation include "warp drive" and the use of singularities in space (that are sometimes known as worm-holes or hyperspace). Serious scientific papers have actually been written on these subjects, but they remain highly speculative, technologically speaking, since they involve untested scientific principles. We shall in this book look at technologies that employ tested physical principles, although relevant practical technological applications are yet to be achieved for those physical laws.

There might be those who would wonder what would happen if we accelerated the ship to near the speed of light. Wouldn't the ship go beyond the speed of light, if we gave it a little more push? The problem is that the ship's velocity is not linearly additive; at almost the speed of light, the combination of the relativistic time dilation effect and the shortening of distances makes the added acceleration diminishingly infinitesimal. If you go through the relatively simple calculations using the famous Einsteinian equations regarding the relativistic time and length, you will find that the added velocity approaches zero as the ship's velocity approaches the speed of light. An infinitely shortened distance divided by an infinitely lengthened time would naturally approach zero – and the added velocity would in consequence approach zero.

What would happen if you launched another ship from a ship that is traveling at almost the speed of light? Certainly, the second ship could be accelerated to almost the speed of light relative to the first ship. However, when measured from an outside frame of reference (e.g., Earth), the velocity of the second ship does not and cannot reach or exceed the speed of light because of the aforementioned time dilation effect and the shortening of distances due to relativity. Please note that the speed of light is not exactly a physical barrier for space ships as is perceived by some people.

[IV] A Single Versus Multiple Interstellar Ships Journeying to the Same Destination

Instead of sending one ship (which is self-sufficient for human survival for many centuries), it might make sense to send at least two ships, possibly more, together as an insurance against unanticipated problems. If we send only one interstellar ship, if a catastrophe should strike it, the nearest help may be too far to do any good. On the other hand, if there is a companion ship traveling close by, a serious mishap on one ship need not be fatal to the mission.

The multiple ships could be designed to be self-sufficient by themselves. Alternatively, a manned ship could be accompanied by an unmanned vessel, which could carry provisions for the first ship but could be converted to serve as a habitat in case the first ship should become unsuitable for human habitation.

[V] Anthropological, Genetic and Other Related Non-Engineering Issues Involved

Traveling interstellar distances would involve not only the purely technological issues of traversing the great distances in space ships; just as importantly, we need most carefully to consider what sort of crew we need to journey to those distant planets. A multi-generation space ship would need to have a sufficient gene pool for the crew to survive and thrive for generations and a social structure that would make it possible for the crew to remain a viable social group. The importance of addressing such issues in a scientific and rational fashion must be emphasized, especially to those whose training has not included anthropology and related subjects. The articles by John Moore, Dennis O'Rourke, and Sarah Thomason address those crucial issues.

[VI] Confronting Unsubstantiated Beliefs and Misconceptions About Interstellar Travel (That are Not Based on Scientific Considerations)

(A) Anthropology, genetics and other related issues

There are physical scientists, hopefully a minority, who are surprised that anthropology and the social sciences in general are important disciplines in planning interstellar flight. Actually, these fields are quite important in making plans for interplanetary voyages. It is important to know the smallest viable number for the crew, their gender and genetic makeup, and their social organization from serious anthropological research. Anthropologists have been studying related issues carefully, examining, for example, the number and the social structure of the Polynesian people in island-to-island migrations in the vast Pacific Ocean in bygone centuries.

Some would like to think that the human race will have learned to live in peace by the time interstellar multi-generation space ships could be built and launched; people aboard the interstellar ships should then be able to live peacefully with each other for countless generations. I am not sure if the human race will have learned to live in peace in several centuries, although it certainly would be wonderful if that should come to pass. But, even if the human race should learn to live in peace on Earth in that not-so-remote future, that wisdom and custom would not automatically be inherited by those people on the ship for numerous generations to come. History shows examples of nations that have learned to form a democratic society, only to lose it in a few centuries. Consider ancient Greece or Rome, as examples.

Careful scientific examinations of the human race, as is customarily done by serious anthropologists and researchers in other related disciplines, must be very much an important and essential part of making plans for an interstellar journey lasting a number of generations.

(B) Science and technology related issues

There are also a number of people (including some physical scientists) who maintain that interstellar travel is impossible because they cannot imagine how it can be achieved technologically.   Some would say that because no ETI (extra-terrestrial intelligence) space ships have visited us, interstellar travel must be impossible. Their reasoning appears to be that intelligent beings must have evolved on thousands, if not millions, of other planets in our galaxy, and it is inevitable that they would visit us if interstellar travel was possible. The absence of such ETI contact demonstrates, they say, that interstellar travel is impossible.

At the beginning of the twentieth century, Simon Newcomb, one of the scientific savants of the time, was asked by a news reporter if a flying machine could actually be built. He answered that it was mathematically impossible for a heavier than air machine to fly in the air. A short while later, the Wright brothers had their first successful flight at Kitty Hawk. The same reporter went back to Simon Newcomb and asked what Newcomb's views might be in light of that success. He answered, in effect, that it was marginally possible for a heavier-than-air machine to fly but a flying machine could not possibly carry a passenger or a cargo. Not long after that, airplanes were being used in World War I; in World War II, airplanes became decisive elements in winning – or losing – the war (although those may not be the finest examples of human inventiveness). We couldn't even imagine now what the world would be like without airplanes and the national and worldwide air transportation systems that airplanes make possible.

When President John F. Kennedy announced his plans in May 1961 to send Americans to the Moon and bring them back safely by the end of the decade, there were many technically trained people who did not think it could be done, since such a feat had never been accomplished before. To those people, what has not been done could not be done.

There are also those who would extrapolate the progress of science and technology only linearly. For example, one such person compared the maximum traveling velocity for a human being in the 18th century (riding on a horseback at 40 kilometers an hour) and the maximum velocity attained in the late 1960s (by an astronaut aboard the Apollo spaceship going to the Moon); he then extrapolated the "rate of progress" to the 26th century, and concluded that the space ship of the 26th century would travel at a few percent the speed of light. Would such an extrapolation be valid? If one took the maximum velocity of a person on a horseback at the time of the Persian Empire several millennia ago and compared it with that of a person on a horseback in the eighteenth century, it would still be the same 40 kilometers an hour. If one applied the rate of change in the maximum velocity over the few millennia preceding the 18th century – which was nil – and extrapolated it to the 26th century (or to any other future century), the maximum velocity for a human being in the 26th century would be 40 kilometers an hour – still riding on that horseback, of course.

It is important to note that scientific and technological progress are usually non-linear – often startlingly so. The science and technology of the future that would be available for building interstellar space ships would indeed be quite different from those imagined from the perspectives of the early twenty-first century. Nevertheless, it is a good idea to start thinking what it will take to mount such an undertaking for the reasons discussed before, and also so that we can begin exploring various scientific and engineering possibilities now – rather than waiting endlessly for "the right time" to come.

*****

Instead of blindly following popular preconceptions and biases about matters that we have not yet had the chance to test or verify, we shall examine in this volume our current state of knowledge, as well as our present state of ignorance, on subjects related to interstellar travel, and we shall see where that will take us.

As a final chapter in this book, we are very pleased to include a thought-provoking article by Freeman Dyson titled *Looking for Life in Unlikely Places*, which is based on a lecture he gave at the Jet Propulsion Laboratory. Section 4 of his essay contains a discussion of interstellar travel by life forms – without the use of space ships.

Acknowledgment:

It gives us pleasure to acknowledge the support we received from Dr. Edward J. Weiler, Dr. Guenter R. Riegler, and Dr. John J. Hillman of the NASA Headquarters in holding this AAAS symposium on *Interstellar Travel and Multi-Generation Space Ships*.

— PART I —

Physical Sciences & Technology Section

Fly Me to the Stars:
Interstellar Travel in Fact and Fiction

Charles Sheffield, Ph.D.
Earth Satellite Corporation
6011 Executive Boulevard
Rockville MD 20852

Abstract

This is a historical survey of concepts of travel beyond Earth, and particularly interstellar travel, beginning at the time when humans first realized that the stars were more than points of light in a crystal sphere.

We distinguish three phases: before 1838, when the distance to a star was first measured; from 1838 to 1905, the year in which Einstein published the papers on relativity that established the speed of light as a limiting velocity for all signal transmission and all material travel; and from 1905 to the present day, when the idea of travel to the stars became of greater interest in both fact and fiction.

The year 1600 seems like a good place to begin. That was when Giordano Bruno was gagged and burned at the stake for proclaiming his belief that the universe contained an infinity of worlds, and that many of those worlds were populated by living creatures. As he wrote, "There is not merely one world, one Earth, one Sun, but as many worlds as we see bright lights around us." Bruno died about a decade before Galileo turned his home-made telescope to the skies, vastly increased the number of "bright lights" in the heavens, discovered mountain ranges on the Moon and four satellites of Jupiter revolving about their primary like a miniature solar system, and changed human ideas forever about the nature of the universe.

Let us summarize the situation in 1600. It had been more than half a century since Copernicus, in 1543, published his great work *De Revolutionibus Orbium Coelestium* (*On the Revolution of Celestial Bodies*) proposing the Sun as the center of the solar system. However, the Ptolemaic Earth-centered view still prevailed in most people's minds, despite Copernicus and the earlier heliocentric theory of Aristarchus of Samos, eighteen centuries earlier. In 1616, Copernicus' work was placed on the Roman Catholic Church's list of "banned books," where it remained until 1835. Most authors would be delighted for their works to remain of interest for so long. However, given the fate of Copernicus' book and Giordano Bruno's frightful end, it would not be surprising if speculations about life on other worlds had gone into a sharp decline after 1616.

In fact, they did not disappear. Johannes Kepler, in 1634, wrote a strange work of fiction entitled *Somnium* – (*The Sleep*). He imagined travel to and life on the Moon. A few years later, in 1638, Francis Godwin wrote *The Man in the Moone*. However, travel to more distant worlds, and in particular travel to the stars, was another matter. No matter what Giordano Bruno had thought and said, to most people the stars were mere points of light, fixed permanently and unmoving within the great crystal sphere that lay beyond all planets. In 1600, Shakespeare could still write, in *Hamlet*, "Doubt thou the stars are fire, doubt thou the Sun does move," without being afraid of questions from his audience.

Before interstellar travel could become a subject for speculation, the stars themselves had to become objects of substance. The vast number of stars was realized when Galileo's telescope was able, about 1610, to decompose parts of the diffuse band of the Milky Way into individual stars. However, the number of those stars, and a plan for the general structure of the Galaxy, had to wait another century and a half until William Herschel had instruments of sufficient resolving power to make detailed star counts. Even then, stellar distances remained a mystery. All that anyone could say was that they were too far away to measure. It was not until 1838 that Friedrich Wilhelm Bessel was able to make a parallax measurement on the star 61 Cygni, based on the annual movement of the Earth around the Sun. That parallax was tiny, just thirty-one hundredths of a second of arc. From it the vast distance to even the nearest stars was established. And from the observed brightness and the estimated distance, astronomers confirmed that stars were objects comparable in size with our own Sun. Evidence of compositions similar to our Sun came after 1860, when, following the invention of the spectroscope, Gustav Kirchhoff formulated general laws for the interpretation of spectra.

Interstellar distances are perhaps still under-appreciated by the general public. They can be illustrated by a simple example. Suppose that a space transportation system was developed able to carry a spacecraft and its crew to the Moon in one minute. Anyone interested in solar system development would drool at the very thought of such a device. That spacecraft would, for reasons to be described later, take a *minimum* of 190 years to reach even the nearest stellar system, Alpha Centauri. Many of the best-known stars are much farther away.

By the mid-nineteenth century, then, the distances facing a would-be interstellar traveler were becoming known. However, it was not at all clear that there was anywhere for that traveler to go. No one had evidence of extrasolar planets, and the argument for their existence was based only on analogy with the structure of the solar system. Some of the theories for planetary formation were not encouraging. One argument, made by James Jeans in 1916 and subsequently, proposed that the planets were created when another star made a close approach to the Sun. The passer-by drew from the Sun a broad filament of matter that condensed to form the planets. This suggested a relatively rare event, occurring according to Jeans's calculations to one star in 4,000 every ten billion years. This would imply a paucity of extrasolar planets. Earlier, in the 1790's, Pierre Laplace had proposed that the planets, together with the Sun, had condensed from a large body of gas and dust. However, James Jeans had published mathematical objections to that notion, and the tidal origin theory was still acceptable in the 1940's.

Moreover, in 1905 Albert Einstein had provided a theoretical problem to go with the practical one of great distance. The theory of relativity asserts that no material object can be accelerated to travel as fast as light. If something is to be sent to the stars, no matter whether it be as small as a grain of sand or as big as a spaceship containing hundreds or thousands of humans, the travel time must at the very least be measured in years, and more likely in centuries.

A few possible destinations of interest, and their associated travel times, are given in Table 1. Note that these are *absolute minimum time values*. They assume that a ship will be able to travel at very close to the speed of light, and attain that speed in a small fraction of total trip time. Actual times would undoubtedly be far longer, and we will look at them later.

Table 1 – Minimal travel times for selected target stars

Destination	Travel time in years
Alpha Centauri	4.4
Sirius	8.7
61 Cygni	11.2
Procyon	11.6
Altair	16.6
Eta Cassiopeia	19.2
Betelgeuse	520.
Galactic center	30,000.
Andromeda Nebula	2,000,000.

A side comment is appropriate here. The times in Table 1 are as measured on Earth, or at the destination. In partial compensation for its strict rule on maximum possible speed, relativity permits the travel times, as measured on board a star ship, to be less. This phenomenon, quite real and experimentally verified, is known as relativistic time dilation. It becomes a significant factor only when a ship is traveling very fast. At 99% of light-speed, onboard time passes seven times as slowly as external time.

During the 19th century, neither the enormity of interstellar distances nor the restrictions imposed by relativity had been a problem for writers of fiction. Jules Verne had found the Moon to be a sufficiently remote target to evoke a sense of wonder. He didn't need the stars to amaze his readers in 1865, when he published *From the Earth to the Moon*. Nor did H.G. Wells, when in 1898, in *The War of the Worlds*, he shifted the focus to Mars, and had the Martians visit us rather than vice versa. For another thirty years after that, the planets of our own solar system provided a big enough playground for the writers of adventure stories. It helped a lot that we didn't know what those planets were like. A habitable swampy Venus was possible, beneath all those clouds – which we now know are clouds of sulfuric acid. As for Mars, writers took it for granted that you could live there. Hadn't Percival Lowell, at the beginning of the century, seen – or, as we now know, imagined – a great system of canals, bringing seasonal snow-melt down from the polar caps for irrigation?

However, by the 1930's the writers of fiction were seeking new places for their heroes and heroines to go. The worlds of the solar system, while still largely unknown in any detail before the space probes of the 1960's and later, offered limited scope. Planets around other stars – lots of planets – were needed. So they were simply assumed to be there. The leading pioneer was Edward E. ("Doc") Smith, who wrote *The Skylark of Space* in 1928 and followed it with many other galaxy-spanning novels. He took it for granted that most stars had planets. More than that, he assumed they were planets habitable by humans, and he applied that assumption with great gusto. So did Ed Hamilton and Jack Williamson, other prolific writers of the same period. (And later, Williamson, whose first story was published in 1928, in 2001 won science fiction's Hugo award for a new story, *The Ultimate Earth*; such a record of sustained production is unmatched in the field).

It took another sixty years before evidence of extrasolar planets was found, to support the wishful thinking of the late 1920's. There had been one apparently successful attempt made at planet detection in the 1940's to 1960's, at Swarthmore's Sproul Observatory. It led to the conclusion that Barnard's Star, which at less than two parsecs distance is the second closest stellar system to Sol, had a planetary companion. Unfortunately, the observations turned out to be caused by changes in the measuring equipment itself, and we had to wait another thirty years until the first good evidence for an extrasolar planet was produced. In 1995, Michel Mayor and Didier Queloz of the Geneva Observatory in Switzerland announced the discovery of a Jupiter-sized planet in orbit around the star 51 Pegasi. This was confirmed within a few weeks by Geoff Marcy and Paul Butler of San Francisco State University, who shortly thereafter discovered two more planets, one orbiting 47 Ursa Majoris and the other orbiting 70 Virginis. Since then, more and more extrasolar planets have been found, bringing the total up close to a hundred.

Even now, discovery of extrasolar planets is far from easy. Planets around other stars do not reflect enough light to be visible from the solar system, and what light they do give off is overwhelmed by that coming from their primary. The detection technique depends on the gravitational perturbation of the parent star by the orbiting planet, and since the mass of the latter is small compared with the mass of the former, observing and measuring a perturbation is a tricky business. In principle, we might observe a change in the star's apparent position in the sky during the period of the planetary year, as star and planet revolve about their common center of gravity. This was the method used in the effort at the Sproul Observatory in the 1960's. It had problems then, and to date it has had no confirmed successes.

The most successful method of detection also relies on the fact that the star and planet orbit around each other, but in this case we look for a periodic shift in the wavelength of the light that we receive. When the star is approaching, its light will be wavelength-shifted toward the blue. When the star is moving away, the light will be shifted toward the red. The tiny difference between these two cases allows us, from the wavelength changes in the star's light, to infer the existence of a planet in orbit around it.

Since both methods of detection depend for their success on the planet's mass being an appreciable fraction of the star's mass, it is no surprise that we have been able to detect only the existence of massive planets, Jupiter-sized or bigger. Many of these orbit much closer to their parent star than the distance of Jupiter from the Sun. While the evidence for them remains annoyingly inferential, we should note that Jupiter-sized bodies closer to their primary than one astronomical unit make it unlikely that Earth-like planets could exist there.

Recently we have seen examples of a more direct method of detection. When a planet moves across the face of its parent star, it occults a small fraction of the stellar disk. The reduction in total light received at our observing instruments is very small, but not too small to be measured. A possible planet orbiting CM Draconis was found in this way. Again, the discovered planet is Jupiter-sized. Similar detection of an Earth-sized planet remains a task for future instruments. The size distribution of planets around other stars remains an open question. However, as the number of extrasolar planets steadily increases, it seems as though a substantial fraction of stars may have planets orbiting them. Edward "Doc" Smith, we might argue, was ahead of his time in both the enormous scope of his stories and in his assumption of planets scattered thickly across the galaxy.

In another way, however, Doc Smith and his contemporaries were very much of their time or even behind it. Rather than acknowledging the limitations implied by the theory of relativity, they chose to ignore it. An imaginative writer such as Jules Verne might reasonably have presumed in the 1860's through 1890's that a spacecraft equipped with some unspecified propulsion system would accelerate to higher and higher speeds, without limit. The stars were distant, but the travel times to them might not be unendurably long. After 1905, such assumptions were unacceptable.

They are unacceptable today, but many writers chafe at the imposition of reality. They dismiss or ignore Einstein's results, invoke an unexplained faster-than-light drive, and go anywhere they choose, as fast as they choose. By popping into an ansible, a Bose node, a portal, a star gate, or some other invented term, travelers arrive where they want to be in nothing flat. The assumption, implicit or explicit, is that there is some way to get around the relativistic limitation imposed by Einstein's work. Perhaps the key is through the multiple connectivity of space-time; perhaps it comes through something much in vogue at the moment, quantum entanglement, which shows that the universe is non-local and influences can travel between coupled particles at faster than the speed of light. Perhaps it employs wormholes, which is one example of multiple connectivity.

All these are convenient devices for story telling. However, in terms of real physics, they are just one step removed from sprinkling the top of your head with fairy dust and wishing you were somewhere else.

By the 1940's, certain writers with more respect for science looked for non-magical solutions to the problem of interstellar travel. At the time, there was one obvious answer. If travel to the stars implied travel times of hundreds or thousands of years, and if the average human lifetime remained less than a century, one must build flying worlds. These would be self-contained biospheres, able to provide their own power, grow their own food, and recycle their own wastes over many centuries. Within them would exist human societies in microcosm, people living their whole lives aboard the ship and finding nothing unnatural in this – any more than we find it unnatural to spend our whole life moving through space within the particular biosphere of the Earth.

There is some debate about who first conceived of and wrote about multi-generation star ships. Credit generally goes to the great Russian space pioneer, Konstantin Tsiolkovsky. He was certainly not writing fiction when he proposed the idea in an essay published in 1928. It is not known when he first thought of the concept. In fiction, Robert Heinlein published the short stories *Universe* and *Common Sense* in 1941, both taking place aboard a multi-generation star ship. Brian Aldiss published a novel on the same theme, *Nonstop*, in 1958. It is interesting that in both the Heinlein and the Aldiss work, the crews of the multi-generation ships have forgotten their own origins – something which seems highly improbable, but provides good story material. The same idea of the multi-generation ship regarded as the whole world can be found in a 1953 story, *Spacebred Generations*, by Clifford Simak.

Multi-generation star ships continued to be written about, but by the 1960's a new idea was in the air – or perhaps we should say, in space. The speed of light is an absolute constant, with a fixed value in vacuum. Given our inability to speed up the speed of light, why not slow down the people? Suppose it takes the best star ship that we can make fifty years to reach Sirius. That is a more than respectable speed, since Sirius is more than two parsecs away. But suppose that instead of going to bed and sleeping for the usual 8 hours, you went to bed and slept for 50 years. You couldn't do this with ordinary sleep, of course, since you would die of starvation in the first few weeks. What you need is "cold sleep," in which the body temperature has been lowered dramatically and body metabolic processes proceed imperceptibly slowly. As a device, cold sleep, or "cryo-sleep" as it is often called, has been even more popular than multi-generation star ships. The first example was probably Don Wilcox's 1940 story, *The Voyage that Lasted Six Hundred Years*. In this curiously improbable tale, the star ship's captain is the only one who hibernates. He wakes up once a century to see how things are going among the non-hibernating crew, and doesn't like what he finds. Social degeneration and disease are the order of the day.

More recently, cryo-sleep was used famously in the movies *2001* and in *Alien*, and in a thousand other stories. The danger with long-term hibernation, however achieved, was pointed out in an A.E. van Vogt story of 1944, *Far Centaurus*. While the cold-sleeping crew are on the way, someone invents a method of faster-than-light travel – and is waiting to meet the travelers when they wake up at their destination.

One practical objection that might be raised to cryo-sleep is that we do not yet know how to put someone into cold sleep and then wake them. We do know, in principle, how to make a self-contained biosphere. However, it would be unwise to be too optimistic about the latter. Experiments so far with self-contained biospheres show that it is difficult to keep them going for more than a few years. This is not good if you are already outside the solar system and headed for Canopus.

Now let us examine the problem from a different perspective. Rather than the minimal travel times of Table 1, what are the *realistic* travel times to the stars, using any method based on today's physics and astronomy? Table 1 assumed that one could travel indefinitely close to the speed of light. A much more likely maximum is a modest fraction of that, perhaps five percent of light-speed. If this seems too slow, let us note that a ship going this fast would travel the distance from Earth to Moon in less than half a minute. It is highly unlikely that a multi-generation ship would attain anything near this speed.

Note, by the way, that our first unmanned star ships have already left the solar system. They are in Table 2, with the launch date of each, the closest stellar target star, and the estimated travel time in years to that star.

Table 2 – Travel Times to Nearby Stars by 20th Century Spacecraft

Spacecraft:	Pioneer 10	Pioneer 11	Voyager 1	Voyager 2
Launch date:	03 Mar 1972	05 Apr 1973	20 Aug 1977	05 Sep 1977
Star:	Ross 248	AC+79 3888	AC+79 3888	Sirius
Travel Time:	32,600 yr	42,400 yr	40,300 yr	497,000 yr

Travel times for today's star ships are long indeed. Table 3 provides travel times for some future ship moving at 5 percent of light-speed. The table ignores the acceleration and deceleration phases of the trip, which are likely to be a substantial fraction of the whole.

Table 3 – Travel at 5% of the speed of light

Destination	Travel time in years
Alpha Centauri	88.
Sirius	174.
61 Cygni	224.
Procyon	232.
Altair	332.
Eta Cassiopeia	384.
Betelgeuse	10,400.
Galactic center	600,000.
Andromeda Galaxy	40,000,000.

We see that we have minimal realistic travel times of a century or more. This presumes an available energy source many times today's planetary capacity. Far more likely, if we wish to travel to a star system possessing something close to an Earth-like world, we must expect travel times of 500 to 1,000 years. We don't yet know where that nearest Earth-like world is located. Perhaps our travel time will be ten thousand years.

However, let us stick with the smaller numbers, and let us look, not forward but *backward*, 500 and then 1,000 years. And let us suppose that we could ask people of those distant times – to be specific, let's say we ask William the Conqueror and Christopher Columbus – to organize the best minds of their day, and come up with a way to travel to the stars.

Assuming that they could understand the problem, which is extremely unlikely, it would be quite impossible for them to produce a meaningful answer. Our technology would be unknown to them. Worse than that, had we described it, no matter how clearly, the explanations would have been *incomprehensible*. Just as incomprehensible, one suspects, as the science and technology of five or ten centuries from now would be to us, were someone from the future to come along and try to explain it.

Star travel is not impossible. It may not even be difficult. The fact that it seems difficult or perhaps impossible to us, today, is merely a measure of the state of our science and technology. This is only discouraging if you say, "I want to see it *now*" – meaning, "I want to see it happen within my own lifetime."

My feeling is that for humans to travel to the stars, it is not necessary for us to sprinkle fairy dust on our heads. Nor do we have to become super-human. We merely need to be patient.

It seems appropriate to end with a quotation from one of the other great pioneers of space research, Robert Goddard. In 1932, in a letter to H.G. Wells, he wrote as follows: "There can be no thought of finishing, for aiming at the stars, both literally and figuratively, is a problem to occupy generations, so that no matter how much progress one makes, there is always the thrill of just beginning."

Now, seventy years later, we too can experience the thrill of just beginning.

The following article by Bob Forward entitled, *Ad Astra*, was originally published in the *Journal of the British Interplanetary Society* (Vol. 49, pp. 23-32, 1996). It is reprinted here with the gracious permission of the *J.B.I.S.*

Ad Astra!

Dr. Robert L. Forward
Forward Unlimited
PO Box 2783
Malibu, CA 90265, USA.

A broad survey of the present status of various concepts for interstellar transport is presented. A conclusion of the paper is that robotic interstellar flight is difficult and expensive, but not impossible. A list of problems in the field needing further study is included.

1. Introduction

It was only a few centuries ago that people began to realize that those points of light in the night sky were suns; like our Sun; and like our Sun, they might have planets around them. Many visionaries then dreamed and wrote of visiting those other planets in ships that traveled between the stars. Later, when astronomers were able to estimate the distance to the nearer stars, others concluded that, because interstellar distances were so immense and human life so short, interstellar travel was impossible.

Travel to the stars will be difficult and expensive. It will take decades of time, gigawatts of power, kilograms of mass-energy and trillions of dollars. Recently, however, some new technologies have emerged and are under development, for other purposes, that show promise of providing propulsion systems that will make interstellar travel feasible within the foreseeable future – if the world community decides to direct its energies and resources in that direction. Make no mistake – interstellar travel will always be difficult and expensive, but it can no longer be considered impossible.

Why should we bother going to the stars if it is so difficult and expensive?

First, there is the visceral explorer's or mountain climber's motive – one goes to see the stars and walk upon the planets, "because they are there." Second, there is the primordial "survival-of-the-species" motive. If the race of human beings is to survive for more than a few millennia, a small portion of the species must leave this fragile blue egg and its inconstant source of heat and travel to another world around another sun, to start a new settlement safe from what might happen to this world or this Sun. Third, and more immediately profitable, would be to find other life forms, hopefully intelligent life forms. Any form of alien life, no matter how simple, will stretch the theories of life held by our biologists and medical doctors, leading them to new ideas, new products and new treatments. The finding of intelligent life will be even more valuable.

Some argue that if intelligent alien life forms existed on other planets, people could communicate with them by radio signals and there is no need to go there in starships. SETI (Search for Extra-Terrestrial Intelligence) projects are presently under way around the world and, although no confirmed signals have been found to date, the search is valuable and should continue. But not all intelligent life forms will have radio. For example, life could evolve on an ocean-covered world to produce intelligent whale-like or octopus-like creatures. Such beings could be highly advanced in music, mathematics, philosophy, hydrodynamics, acoustics and biology, but they would have no technology based on fire or electricity. To interact with those types of intelligent beings we must go there in starships and visit them.

2. Interstellar Distances

It is not easy to comprehend the distances involved in interstellar travel. Of the billions of people living today on this globe, many have never traveled more than 40 kilometers from their place of birth. Of these billions, a few dozen have traveled to the Moon which, at a distance of almost 400,000 kilometers, is ten thousand times 40 kilometers away. Recently an interplanetary probe passed the orbit of Neptune, more than ten thousand times further out at over 4,000,000,000 kilometers. The nearest star is at 4.3 light years, 10,000 times further away than Neptune. Yet, within 20 light years from the Sun, there are 59 stellar systems containing 81 visible stars – plenty of targets for us to explore while engineers are developing even more powerful interstellar transport systems. The prime targets for interstellar missions are listed in the following Table [1].

Table 1: Prime Targets for Interstellar Missions

Stellar System	Distance (ly)	Remarks
Alpha Centauri	4.3	Closest system. Three stars, (G2, K5, M5). Component A almost identical to Sun (a G2 star).
Barnard's Star	6.0	Small, low luminosity M5 red dwarf. Next closest to Solar System.
Sirius	8.7	Large, very bright A1 star with a white dwarf companion.
Epsilon Eridani	10.8	Single K2 star slightly smaller and cooler than the Sun. May have solar system type planetary system.
Tau Ceti	11.8	Single G8 star similar to the Sun. High probability of possessing a solar system-type planetary system.

3. Time, Acceleration, Velocity and Energy Requirements

The nearest star system is 4.3 light years away, so even if a magical method could be found to rapidly accelerate a vehicle up to close to the speed of light, it would take a minimum of 4.3 years before the vehicle arrived at Alpha Centauri, and another 4.3 years before the information about what was found there was returned to Earth, either by radio or by a returning vehicle. Thus, the minimum time for an interstellar mission is 8.6 years, much longer than most of the interplanetary missions presently being carried out. The first real interstellar missions, probably carried out by robotic probes, will require time to accelerate up to speed and will use sub-relativistic cruise speeds, so they will necessarily take many decades to accomplish. Yet, if a mission takes too long, then it is likely that the spacecraft will be passed, by a follow-on mission launched decades later with a better propulsion system. Thus, unless an interstellar mission can be completed in less than 50 years, it probably should not be started. Instead, the money should be invested in designing a better propulsion system. As a result, missions to the nearest stars are required to have mission times between 8.6 and 50 years.

A mission time requirement of 50 years produces acceleration and velocity requirements. If a vehicle accelerates at one Earth gravity (1 gee) for one year, which would require a propulsion system better than any presently imagined, it will have reached greater than 90% of the speed of light in a distance of only 0.5 light years. It could now coast for 3 years, then decelerate for one year to arrive in Alpha Centauri in about 5 years. This time is not much longer than the minimum travel time of 4.3 years. Therefore, vehicle accelerations greater than 1 gee do not improve the mission time to the nearest stars significantly. Accelerations less than 0.01 gee result in mission times that are too long. A vehicle with an acceleration of 0.01 gee would take 20 years to reach the Alpha Centauri half-way point, and would have reached only 20% of the speed of light. It would then take 20 years to decelerate. Add in the 4.3 years of communication time and the mission approaches the 50 year maximum time requirement. Thus, missions to the nearest stars require accelerations between 0.01 and 1.0 gee.

A time requirement of less than 50 years and an acceleration requirement of between 0.01 and 1.0 gee, produces requirements on the cruise velocity of the vehicle. Tables II and III show the range of mission times of a one-way flyby probe to the nearest star and a distant star, assuming an acceleration capability of 0.1 gee to various maximum cruise velocities.

Table II: Mission Times To Alpha Centauri at 4.3 light years.

* c is the speed of light

Maximum Velocity (c)	Acceleration Time-0.1 gee (years)	Coast Time (years)	Data Return (years)	Mission Time (years)
1.00	10	0	4.3	14
0.50	5	6	4.3	15
0.40	4	9	4.3	17
0.30	3	13	4.3	20
0.20	2	21	4.3	27
0.10	1	43	4.3	48
0.05	0.5	85	4.3	90

Table III: Mission Times to Tau Ceti at 11.8 light years.

* c is the speed of light

Maximum Velocity (c)	Acceleration Time-0.1 gee (years)	Coast Time (years)	Data Return (years)	Mission Time (years)
1.00	10	6	12	28
0.50	5	21	12	38
0.40	4	28	12	44
0.30	3	40	12	55
0.20	2	60	12	74
0.10	1	120	12	133

The nearest star to us is Proxima Centauri, a small red dwarf star that is part of the three-star Alpha Centauri system. The primary star in the Alpha Centauri system, Alpha Centauri A, is similar to our Sun. Also known as "Rigil Kent," it is the brightest object in the southern constellation of Centaurus and the third brightest object in the sky after Sirius and Canopus. To carry out even a one-way unmanned probe mission to the Alpha Centauri system in the 50-year professional lifetime of the humans that launched the probe will require a minimum velocity of 10% of the speed of light. At that speed it will take the probe 43 years to get there and 4.3 for the information to get back to us.

Further away are other single stars like our Sun that seem the best candidates for finding an Earth-like planet. These stars are Epsilon Eridani at 10.8 light years and Tau Ceti at 11.8 light years. To reach these in a reasonable time will require starship velocities of 30% of the speed of light. At this speed, it will take nearly 40 years to travel the 11-12 light years, plus another 11-12 years for the information to return to Earth.

Although we need to exceed 10% of the speed of light to get to any star in a reasonable time, if we can attain a cruise velocity of only 30% of the speed of light, then there are 18 star systems with 26 visible stars and hundreds of possible planets with 12 light years of the Sun. This many stars and planets within reach, at 30% of the speed of light, should keep us busy exploring while engineers develop faster starship designs.

As can be seen in the tables, as the probe coast velocity becomes higher, the coast time becomes shorter, but the acceleration periods become longer. Increasing the coast velocity does not significantly improve the mission times to the nearest stars. For missions to the nearest stars the required cruise velocities are between 0.1 and 0.5 c.

The important conclusion is that if the mission time is constrained to less than 50 years, then a usable interstellar probe destined for the nearest stars will likely operate in the acceleration range from 0.01 gee to 1.0 gee (typically 0.1 gee or 1 m/s^2), and the velocity range from 0.1 to 0.5 c (typically $c/3$ or 10^8 m/s). At these speeds, relativistic effects will be measurable, but not large. For v = $c/3$, the time dilation factor is only 5.7 percent (Eq. 1 in side bar part A on page 34).

The energy requirements for an interstellar probe will be high. Assuming a minimal flyby vehicle massing 1 metric ton (1,000 kg = 1 Mg) with a coast velocity of $c/3$, the kinetic energy in the vehicle is 5 x 10^{18} J, about equal to the present energy consumption of the U.S.A. for three weeks. This amount of energy masses 56 kg, so the total mass of the vehicle is really 1,056 kg, not 1,000 kg. The actual energy requirements for the mission will depend upon the efficiency of the propulsion system in converting input energy into kinetic energy of the vehicle.

With the minimum energy requirements determined, the power requirements can be calculated. Assuming an acceleration period of 3.3 years (producing a coast velocity of $c/3$ at an acceleration of 0.1 gee), the minimum propulsion power that must be applied to push 1 Mg to $c/3$ at 0.1 gee is 50 GW. This is a remarkably low value. The three Space Shuttle Main Engines operate at a power of 22 GW, so rocket engines of comparable power and energy density to that required for interstellar flight have been built and flown. If rockets are used, however, then in order to keep the rocket mass ratio reasonable, completely new rocket engine concepts must be developed that have an exhaust velocity comparable to the required mission velocity of $c/3$ = 10^8 m/s or a specific impulse approaching 10^9 sec. Alternatively, non-rocket propulsion systems can be used.

4. The "Impossibility" of Interstellar Flight

There have been "scientific" papers "proving" that interstellar flight is impossible. They make the assumption that rockets are used for a round-trip manned mission to a star at 10 light years distance or more. The spacecraft is assumed to accelerate at 1 gee until turnover, decelerate at 1 gee, and then return in the same manner. Even assuming perfect matter-antimatter engines, when the mass ratio is calculated using the relativistic rocket equation (Eq. 10), the amount of antimatter needed exceeds the mass of the Earth!

There are no flaws in the mathematics of these papers, the flaws are in the assumptions. First, if some non-rocket propulsion technique is used instead of a rocket then the spacecraft does not have to carry its reaction mass or energy source: the relativistic rocket equation does not apply and the fuel mass required does not rise exponentially. Second, the assumption of constant acceleration at 1 gee means that after one year, the rocket is close to the speed of light. Additional acceleration does not increase the speed, or decrease the time of the mission (to those on Earth paying for the mission – the astronauts do benefit from the time dilation effect), it only makes the vehicle and its load of fuel heavier and harder to push! It would be better to optimize the mission so the acceleration and deceleration periods are only a year or so, with most of the mission carried out at a coast velocity somewhat below the speed of light. The energy and mass requirements will still be high but will not be impossible.

5. Relativistic Rocket Mechanics

Travel at significant fractions of the speed of light requires the use of equations for relativistic mechanics instead of classical mechanics[2]. In relativistic mechanics, even the fundamental properties of time, length and mass vary with the relative velocity of the observer. For example, see the time dilation equation in side bar part A. This equation results in the so-called "twin paradox," where a traveling twin, upon taking a round trip journey at relativistic speeds, is found on return to be younger than the stay-at-home twin.

In the same manner, the length of a vehicle moving at relativistic speeds appears to contract relative to its stationary length (see side bar part B).

Similarly, the mass of a vehicle moving at relativistic speeds increases as speed increases (see side bar part C).

Another result of relativistic mechanics is that nothing can go faster than the speed of light (see side bar part D).

[A] The time t measured on a space vehicle moving at a velocity v is shorter than the time t_0 measured by a stationary observer, by the relation:

$$t = \sqrt{1 - \frac{v^2}{c^2}}\, t_o \tag{1}$$

where $c = 299.8 \times 10^6$ m/s = the speed of light.

[B] The length l of a moving space vehicle along the direction of the velocity v, as measured by a stationary observer, is shorter than the length l_0 measured when the vehicle is at rest by:

$$l = \sqrt{1 - \frac{v^2}{c^2}}\, l_o \tag{2}$$

[C] The mass m of a moving space vehicle is greater than the rest mass m_0 of the vehicle by the relation:

$$m = \frac{m_o}{\sqrt{1 - \frac{v^2}{c^2}}} \tag{3}$$

The rocket equation for the mass ratio of a relativistic rocket[3] is different from the classical mass ratio. In the classical derivation, a rocket with an initial mass M ejects an amount dm of reaction mass at an exhaust velocity with respect to the rocket of w, assumed to be a constant. In the center of mass of the system, the resultant velocity of the rocket is U and the velocity of the reaction mass is u.

[D] If a vehicle is moving at a velocity v, and shoots a projectile forward with a velocity w with respect to the moving vehicle, then the velocity u of the projectile as seen by the stationary observer is not $u = v+w$, but instead is:

$$u = \frac{v+w}{1+vw/c^2} \tag{4}$$

This equation always produces a velocity for u that is less than the speed of light.

Classical

$$U \longleftarrow \quad M \xrightarrow{w} \quad dm \longrightarrow u$$

Relativistic

$$\frac{M_o}{\sqrt{1-U^2/C^2}} \qquad \frac{dm_o}{\sqrt{1-u^2/c^2}}$$

In the side bar to the right, the derivation of the classical rocket equation (left set of equations) is compared with the relativistic derivation (right set of equations).

For the relativistic derivations, the masses are replaced with their relativistic equivalents (Eq. 3), which vary with the velocities U and u.

From the law of conservation of mass (mass-energy in the relativistic case):

$$dM = -dm \qquad d\left[\frac{M_o}{\sqrt{1-U^2/c^2}} \cdot c^2\right] = -\frac{dm_o}{\sqrt{1-u^2/c^2}} \cdot c^2 \tag{5}$$

From the law of conservation of momentum:

$$d(MU) = u \cdot dm \qquad d\left[\frac{M_o}{\sqrt{1-U^2/c^2}} \cdot U\right] = u \cdot \frac{dm_o}{\sqrt{1-u^2/c^2}} \tag{6}$$

From the law of addition of velocities (Eq. 5):

$$u = w - U \qquad u = \frac{w-U}{1-wU/c^2} \tag{7}$$

Expanding the derivatives and combining the above equations produces:

$$\frac{dM}{M} = -\frac{dU}{w} \qquad \frac{dM_o}{M_o} = -\frac{dU}{w(1-U^2/c^2)} \tag{8}$$

Integrating this result gives:

$$\ln M = -\frac{U}{w} \qquad \ln M_o = \frac{c}{2w}\ln\left[\frac{c+U}{c-U}\right] \tag{9}$$

If the initial mass of the rocket is M_i and the final mass is M_f, then the rocket mass ratio R $= M_i/M_f$ needed to reach the mission velocity $U = \Delta V$ is

$$R = e^{\frac{\Delta V}{w}} \qquad\qquad R = \left[\frac{1+\Delta V/c}{1-\Delta V/c}\right]^{c/2w} \qquad (10)$$

These are the classical and relativistic rocket equations. For the relativistic case, there is a maximum exhaust velocity for the reaction mass that is given by:

$$w = \left[e(2-e)\right]^{1/2} c \qquad\qquad (11)$$

where e is the fuel mass fraction converted into kinetic energy of the reaction mass.

6. Proposed Methods for Interstellar Flight

There are many different concepts for achieving interstellar flight, each using a different technology. A bibliography of most of the concepts can be found in Category 02 of Forward[4], and updates, while Forward[5], is a review of the more feasible concepts. Some concepts are more applicable to small flyby probes, while others can be considered for a round-trip crewed mission. The first three discussed are rockets that carry along their reaction mass, energy source, and engine. As will be seen, any sort of rocket, even an antimatter rocket, has marginal performance for interstellar missions. Only non-rocket propulsion systems offer any prospects for travel to even the nearest star systems. The most promising concepts involve some sort of beamed power propulsion system, which are non-rocket propulsion systems where the heavy parts of a rocket (the reaction mass, energy source and the "engine" that puts the energy into the reaction mass) are all kept in the Solar System near the Sun where there are large amounts of mass available, and the energy source (usually the abundant sunlight), and the "engine" can be maintained and even upgraded as the mission proceeds.

6.1 Nuclear Electric Rocket

The most advanced form of high specific impulse propulsion available today is electric propulsion, while the highest energy density electric power source available is a space nuclear reactor combined with some sort of thermal-to-electric generator. The Jet Propulsion Laboratory in California, USA carried out a study of an interstellar "precursor" mission that was restricted to the use of tested space technology. Called the

TAU (Thousand Astronomical Unit) mission, its goal was to send instruments to distance of 1,000 A.U., or one thousand times the distance from the Earth to the Sun. This would be 25 times further than the distance to Neptune.

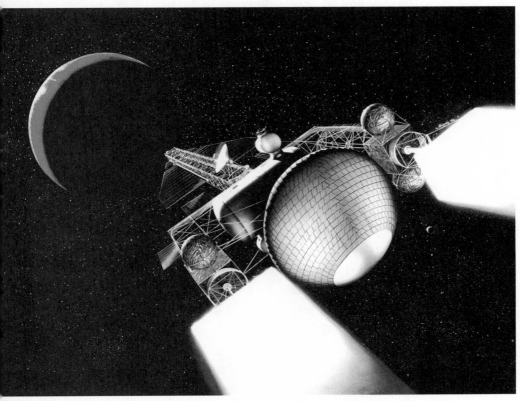

Artist's concept of a nuclear electric-propelled vehicle, about the size of a football field, firing banks of ion thrusters.
Courtesy of NASA

The final spacecraft design[6] was a 60 Mg nuclear electric rocket consisting of a 5 Mg spacecraft, including science instruments, a 1 MW electric nuclear reactor massing 15 Mg to provide the propulsion power, a dozen 12,500 sec specific impulse xenon ion electric thrusters, and 40 Mg of liquid xenon for reaction mass. The TAU spacecraft would be capable of reaching a cruise velocity of 105 km/s (0.00035 c) after burn-up of the nuclear fuel in 10 years, would reach 1,000 AU in about 50 years, and would take 12,500 years to reach Alpha Centauri. Such a spacecraft would be useful for exploring space just beyond the solar system in order to search for planets beyond Pluto and to find nearby "brown dwarfs" – objects bigger than planets but smaller than stars, that glow faintly in the infrared. Since such a spacecraft will take over ten thousand years to reach the nearest stars, it is obviously not suitable for true interstellar travel. Something more powerful is needed.

6.2 Nuclear Pulse Rocket

A faster nuclear-powered interstellar rocket would be one propelled by nuclear bombs. Called the Orion rocket, it was proposed in the late 1950s by Stanislaw Ulam at the Los Alamos National Laboratory in New Mexico, USA. The original goal of the Orion Project was to send manned spacecraft to Mars and Venus. The Orion rocket would be propelled by a series of small nuclear bomblets that explode after they are ejected from the rear of the spacecraft. The hot plasma from the nuclear explosions strikes a "pusher plate," which absorbs the impulse from the explosion and transfers it through large "shock absorbers" to provide a relatively smooth thrust to the main structure. Enough work has been done on the concept, especially the survivability of the pusher plate under repetitive nuclear blast shocks, to determine that nuclear pulse propulsion is a technologically feasible concept. A review and bibliography can be found in Martin and Bond.[7] An alternate design where a "parachute" out in front of the main vehicle replaces the "pusher plate" and elastic "shrouds" act as the "shock absorbers" has been recently proposed.[8]

These ideas for an interplanetary rocket have been extrapolated into a design for a starship.[9] Because of the limitations on the minimum size of the bomblets, which determines the size of the pusher plate, the starship must be quite large, with a payload of 20,000 Mg, enough to support a small town of many hundred crew members. The total mass would be 400,000 Mg, including a fuel supply of 300,000 bomblets weighing about a Mg each. (This is approximately the world's supply of nuclear bombs.) The bombs would be exploded every three seconds, pushing the starship at one Earth gravity. After ten days at 1 gee, the Orion rocket would reach a velocity of 3% of the speed of light. At 3% of the speed of light, Orion would need 140 years to reach Alpha Centauri, and more than 300 years to get to the more interesting star systems such as Tau Ceti and Epsilon Eridani. To decelerate at the target star, it would need to be redesigned to have two stages with the first stage massing 1,600,000 Mg. Alternatively, some sort of "drag brake," such as a magnetic sail, would have to be invented to slow it down at the other end.

Because these travel times are so long, the habitat in the starship would have to be a self-contained world with all of the amenities of life on Earth. Although Orion has minimal performance for the task of interstellar travel, it is one concept for a starship that could have been built and launched on its way twenty years ago.

The Dyson *Orion* interstellar rocket, a 400,000 ton spaceship with a payload of 20,000 tons,
which could accommodate a crew of hundreds, would be propelled by 300,000 nuclear
bombs of one ton each. The bomb explosions, once every three seconds, would produce
a 1 gee acceleration. After ten days, the *Orion* starship would achieve a velocity of 3%
of the speed of light, flying by Alpha Centauri in 140 years.
Courtesy of NASA

6.3 Antimatter Rockets

Research is just starting on a new nuclear propulsion energy source that is more
than a hundred times as powerful as nuclear energy. This new source of energy is
antimatter. For every elementary particle that makes up atoms, such as the proton,
neutron, and electron, there exists a mirror-image particle, the antiproton, antineutron,
and antielectron – also called the positron. The antiparticles have the same mass as the
normal particles but their electrical charges and magnetic fields are reversed. When an
antiproton is put near a proton, the two attract each other and almost instantly annihilate
one another converting all of the mass of both particles into energy. One third of the
energy appears in the form of gamma rays, while two thirds is converted into charged
particles called pions. Moving at 94% of the speed of light when generated, the pions

exist for enough time to travel a distance of 21 meters (60 feet). This distance is long enough for the kinetic energy of the pion explosion to be redirected by a rocket nozzle made out of magnetic fields into a one-way stream that will provide rocket thrust.

Antiprotons are being made and stored for weeks at a time at the European Organization for Nuclear Research in Switzerland. The present production efficiencies are low but techniques exist to improve them by orders of magnitude.[10,11] Storing the antimatter fuel is relatively easy. Scientists working with atomic and molecular beams have already experimentally demonstrated methods for slowing, cooling and storing atoms without letting them touch matter. They have built atom traps using lasers, electric fields, and magnetic fields. In the coming decades it is likely that a significant amount of antimatter will be produced and stored.

Antimatter Rocket Equation: To those rocket engineers inured to the inevitable rise in vehicle mass ratio with increasing mission difficulty, antimatter rockets provide relief, for the mass ratio of an antimatter rocket for any mission is always less than 4.9:1.[12] In an antimatter rocket, the source of the propulsion energy is separate from the reaction mass. Thus, the total initial mass of the rocket consists of the empty mass of the vehicle, m_v, the mass of the reaction mass, m_r, and the mass of the energy source, m_e, half of which is the mass of the antimatter, m_a. The standard rocket equation (Eq.10) mass ration is now (assuming $m_r >> m_e$):

$$R = e^{\Delta V/v} = \frac{M_i}{M_f} = \frac{m_v + m_r + m_e}{m_v} \approx \frac{m_v + m_r}{m_v} \tag{12}$$

The kinetic energy T in the reaction mass at exhaust velocity v comes from the conversion of the fuel rest-mass energy into thrust with an energy efficiency ϵ:

$$T = \frac{1}{2} m_r v^2 = \epsilon m_e c^2 \tag{13}$$

Solving Eq. 13 for the reaction mass m_r, substituting into Eq. 12, and solving for the mass of the energy source m_e produces the equation:

$$m_e = \frac{m_v v^2}{2\epsilon c^2} (e^{\Delta V/v} - 1) \tag{14}$$

The minimum in the amount of antimatter required to perform a mission with a given ΔV is found by setting the derivative of Eq. 14 with respect to the exhaust velocity v equal to zero, and solving (numerically) for the exhaust velocity:

$$v = 0.63 \Delta V \tag{15}$$

Substituting Eq. 15 into Eq. 12, it is found that since the optimal exhaust velocity is proportional to the mission velocity, the vehicle mass ratio is a constant:

$$R = e^{\Delta V/v} = e^{1.59} = 4.9 = \frac{m_v + m_r}{m_v} \tag{16}$$

where the reaction mass m_r is 3.9 times the vehicle mass m_v, while the antimatter fuel mass is negligible. Amazingly enough, this constant mass ratio is independent of the efficiency ϵ with which the antimatter energy is converted into kinetic energy of the exhaust. (If the antimatter engine is of low efficiency, then more antimatter will be needed to heat the reaction mass to the optimum exhaust velocity. The amount of reaction mass needed, however, remains constant.) Provided antimatter engines are developed that can handle the very high exhaust velocity jets implied by Eq. 15, this constant mass ratio holds for all conceivable missions in the solar system and only starts to deviate significantly for interstellar missions where the mission characteristic velocity starts to approach the speed of light.[13]

The amount of antihydrogen needed for a specific mission is obtained by substituting Eq. 15 into Eq. 14 to get the mass of the energy source m_e. The antimatter needed is just half of this, and is found to be a function of the square of the mission characteristic velocity ΔV^2 (essentially the mission energy), the empty mass of the vehicle m_v, and the conversion efficiency ϵ:

$$m_a = \frac{1}{2} m_e = \frac{0.39 \, \Delta V^2}{\epsilon c^2} m_v \tag{17}$$

Antimatter rockets have been evaluated for interstellar missions.[14,13] The mass ratio required is 5:1 up to a cruise velocity of 0.5 c, then rises slowly to 10:1 at 0.95 c. The amount of antimatter m_a needed for a vehicle of mass M carrying out a mission with characteristic velocity Δv is given by Eq. 17. To send a 1 Mg vehicle on a 48 year flyby mission to Alpha Centauri at 0.1 c, would require four tons of reaction mass and 9 kg of antimatter. The optimum exhaust velocity for this mission (Eq. 15) is 0.063 c (19 Mm/s or a specific impulse of 190 million sec). No engine exists that is capable of handling the high temperature plasma this specific impulse implies, although ideas have been published about various designs that use magnetic fields for reaction chambers and exhaust nozzles.

7. Rocketless Rocketry

Although an antimatter rocket is the ultimate in rockets, it is not necessary to use the rocket principle to build a starship. A typical rocket consists of a payload, some structure, an energy source, some reaction mass (in chemical rockets the reaction mass and energy source are combined together into the "fuel"), and an engine. Because a rocket has to carry its fuel along with it, its performance is significantly limited. The more difficult the mission, the more fuel the rocket must carry. The extra fuel makes the rocket heavier, and so it must carry even more fuel to push that fuel. Soon the rocket is nearly all fuel and can carry only a minuscule payload. It is, however, possible to conceive of starship designs that do not use the rocket principle. One method is to "beam" the fuel to the starship as needed. Another is to have the starship "pick up" its fuel along the way, as in the Interstellar Ramjet.

7.1 Interstellar Ramjet

One of the oldest interstellar transport concepts, and the first of the non-rocket propulsion systems, is the interstellar ramjet.[15 and entries in category of 02.05 RAMJET in Forward 4 and updates] The interstellar ramjet consists of a vehicle carrying the payload, a fusion reactor and a large scoop to collect the hydrogen atoms that exist in space. The hydrogen atoms are used as fuel in the fusion reactor, where the fusion energy is released and turned into kinetic energy of the reaction products, which are exhausted to provide thrust for the vehicle. Alternate versions merely collect the hydrogen atoms for use as reaction mass and heat the hydrogen either by onboard nuclear reactors, stored antimatter, beamed laser power, or beamed antimatter.

The ramjet is the only interstellar transport concept that has the potential to reach ultrarelativistic speeds, since the faster it travels, the more fuel it gathers. As a result, the interstellar ramjet is the only starship concept known that can reach velocities close to that of light, where subjective time aboard the starship flows orders of magnitude more slowly than Earth time. This would allow human crews to travel throughout the galaxy and even between galaxies in a single human lifetime. Unfortunately, there are not physically feasible designs presently known for constructing a lightweight scoop that can collect enough hydrogen to achieve reasonable accelerations and cruise velocities. The primary problem is that any magnetic field configuration that is stronger near the "throat" of the scoop than at the perimeter repels the incoming charged particles rather than scooping them in. The concept of a starship picking up its fuel during its journey through "empty" space is too valuable to be discarded lightly, however, and future scientists and engineers should keep working on the remaining problems of the interstellar ramjet until this concept evolves into a real starship.

The Bussard interstellar ramjet, a non-rocket spaceship that picks up its own fuel from space.
It would carry a large scoop to collect hydrogen atoms that exist in space. The atoms would
be used as fuel for the ship's nuclear fusion engine. If the scoop could be built, it would be
the only starship that could reach velocities close enough to light speed that subjective time
aboard the spacecraft would flow orders of magnitude more slowly than Earth time.

Courtesy of NASA

8. Beamed Power Starships

There is another class of starship that does not have to carry along any energy
source, or reaction mass, or even an engine. Consisting only of payload and structure,
these starships use the technique of beamed power propulsion. In a beamed power
propulsion system the heavy parts of a rocket (the reaction mass, the energy source and
the engine) are all kept at home in the solar system near the Sun, where there is an
unlimited amount of energy available. In addition, the engine can be maintained and
upgraded by engineers back home as the mission proceeds. During operation of the
beamed power propulsion system, the energy source (typically the Sun) supplies power
to the engine (the beaming apparatus) which puts the energy into the reaction mass
(either a beam of matter or electromagnetic energy) which is beamed toward the
starship. The starship structure interacts with the beam (either by capturing or reflecting
the beam) to obtain energy and momentum.

If the beam is made up of electromagnetic protons, then there are certain physical limitations on the beam diameter with distance for a given size of transmitter. A transmitting antenna of electromagnetic radiation of wavelength λ, and diameter D, is able to create a spot of diameter d at a distance s given by

$$Dd = 2.44s\lambda \tag{18}$$

where the 2.44 factor indicates that the spot size diameter d is not measured at the half-power point, but at the first null in the Bessel function for a circular aperture, where 84% of the transmitted power falls. If the collecting aperture on the receiving vehicle equals or exceeds this spot size, then essentially all the transmitted electromagnetic power is collected. If the receiving antenna has a smaller diameter, then the power collected is proportional to the relative areas.

Particle beams have similar divergence equations, which depend uniquely upon the specific piece of transmitting apparatus and the particles being beamed.

8.1 Beamed Particle Propulsion

In the beamed pellet propulsion system invented by Singer[5], small pellets would be accelerated by a long electromagnetic mass driver in the Solar System toward the target star. Riding on the pellet beam would be the interstellar vehicle, which intercepts the pellets and reflects them back using a strong magnetic field. Guiding of the pellets to high absolute accuracy is not needed if the vehicle is designed to be a self-adjusting "beamrider" and the angular variations between successive pellets can be kept small. Instead of having to hit a meter-sized vehicle at interstellar distances, the beam guidance system only needs to insure that the beam passes through the AU-sized planetary system of the target star. Deceleration would be accomplished by rebounding the pellets from an expendable unmanned lead ship to decelerate the main vehicle at the target system.

A similar approach using a beam of charged particles to push on a magnetic sail has recently received a good deal of study.[17,18,19] A magnetic sail is a simple loop of high-temperature superconducting wire carrying a persistent current. The charged particles in the beam are deflected by the magnetic field, producing thrust. Thermal analyses of the radiation cooling of the loop of superconducting wire to the 2.7 K background temperature of space indicate that a properly Sun-shielded cable can be passively maintained at a temperature of 65 K in space, well below the superconducting transition point for many of the new high-temperature superconductors.

Other versions of this type of system involve beaming fuel pellets to the vehicle, which the vehicle collects and uses as an energy source. In some variations, the fuel particles are sent out ahead of time to form a "runway" of fuel that the vehicle collects on its way.

3.2 Beamed Microwave Propulsion

Another type of beamed power propulsion system would use beams of microwave photons to drive the starship. An advantage of microwave energy is that it can be produced and transmitted at extremely high efficiencies. A disadvantage is that is that microwaves are difficult to form into narrow beams that extend over long distances. Because of this short transmission range, the microwave-pushed starship must accelerate very rapidly to reach the high velocities needed for interstellar travel. The accelerations required are larger than a human being can stand, and so microwave-pushed starships seem to be limited to use by robots. One design of such a robotic starship is called Starwisp because of its extremely small mass.[5] Starwisp is an ultralight, high-speed, interstellar flyby robotic probe pushed by beamed microwave power. The basic structure of the starship consists of a 1 km diameter hexagonal wire mesh sail with microcircuits at the intersections of the wires. The microwave energy to power the starship would be generated by a solar power station that would be orbiting the Earth. The microwaves from the station would be formed into a beam by a large segmented lens made of concentric rings of wire mesh separated by spaces.

The microwaves in the beam have a wavelength that is much larger than the openings in the wire mesh of the Starwisp sail. Thus, when the microwave beam strikes the wire mesh, the beam is reflected back. The radiation pressure from the reflected microwave beam gives a push to the sail. If the station is emitting 10 GW of microwave power, the 20 g Starwisp will be accelerated at 115 times Earth gravity. The high acceleration allows the mesh to reach a coast velocity of 20% c while still within the solar system and within range of the transmitting lens.

Prior to Starwisp's arrival at Alpha Centauri 21 years later, the transmitter floods the star system with microwave energy. Using the wires in the mesh sail as microwave antennae, the microcircuits on Starwisp collect the microwave energy to power their optical detectors and logic circuits to form images of the planets in the system. The phase of the incident microwaves is sensed at each point of the mesh and the phase information used to form the mesh into a retrodirective phased array microwave antenna that beams the images back to Earth.

The Forward Starwisp, an ultralight unmanned interstellar flyby probe
that masses only 20 grams. Consisting of a wire mesh with microcircuits at
each intersection of wires, it would be propelled by microwaves generated
by a solar power station in orbit around the Earth (left), and shaped into a
beam by a large segmented lens (center). The microwave beam would
accelerate Starwisp at 115 gees, and it would reach 20% of the speed of
light in two weeks, arriving at Alpha Centauri in 21 years. There, Starwisp
would use its light sensitive microcircuits as an "eye" and send back to
Earth pictures of any planets.

8.3 Beamed Laser Propulsion

In beamed laser propulsion,[16] vehicles with large lightsails are pushed by the
photon pressure from a solar-pumped laser array based in the Solar System. The key
element in the system is a final lens in the laser beaming system that is 1,000 km in
diameter. With this size transmitting aperture, the spot size of the laser beam (Eq. 18
would be 100 km in diameter at the 4.3 light years distance of Alpha Centauri (1/10th the
size of the transmitting aperture) and 1,000 km at 44 light years distance. Thus, all the
transmitted laser power is available at the vehicle at any time during the mission, even
over interstellar distances. The lens would be a Fresnel zone-plate consisting of rings of
1 μm thick plastic alternating with empty rings, with a mass estimated at 560,000 Mg.

For an unmanned flyby probe missions of a 1 Mg vehicle that is roughly one-third
each of structure, sail and payload, the diameter of the sail would be 3.6 km. A laser
power of 65 GW would give the sail an acceleration of 0.036 gees. Maintained for three
years, this would produce a coast velocity of 0.11 c. The probe would fly through Alpha
Centauri 40 years from launch.

Since the photons reflected from the sail still have much of their energy left in them at these sub-relativistic speeds, laser pushed lightsails are only energy efficient at near-relativistic velocities (> 0.5 c) where their frequency is strongly reshifted by the Doppler shift. To improve the energy efficiency at lower speeds, spacecraft have been proposed[20] which absorb the laser light in highly efficient solar cells and use the electricity for an ion drive. This improves the energy efficiency at the cost of a more complex spacecraft structure.

A round-trip manned mission using laser-pushed lightsails would involve larger diameter sails and more massive vehicles that would require larger amounts of laser power. For a round trip to Epsilon Eridani at 11 light years, the diameter of the lightsail would be 1,000 km, the same as the transmitter lens (a larger transmitter lens is not required). The lightsail will be divided into three nested circular segments. The total vehicle mass would be 80,000 Mg including 3,000 Mg for the crew, their habitat, and their exploration vehicles. The lightsail would be accelerated at 0.3 gees by 43,000 TW of laser power. At this acceleration, the vehicle will reach 0.5 c in 1.6 years. The expedition will reach Epsilon Eridani in 20 years Earth time and 17 years crew time. At 0.4 light years from the target star, the 320 km rendezvous portion of the sail is detached from the center of the lightsail and turned to face the 1,000 km diameter ring sail that remains. The laser light from the Solar System reflects from the ring sail, which acts as a mirror. The reflected light decelerates the smaller rendezvous sail and brings it to a halt in the Epsilon Eridani system. The crew then explores the system for a few years using the lightsail as a solar sail. For return, the 100 km diameter return sail is detached from the center and turned to face the 320 km diameter ring sail that remains. Laser light beamed from the solar system reflects from the ring sail onto the 100 km diameter return sail and accelerates it up to speed back toward the Solar System. As the return sail approaches the Solar System 20 Earth-years later, it is brought to a halt by a final burst of beamed laser power. The crew has been away 51 years (5 years exploring) and have aged 46 years.

Beamed laser propulsion seems to be the best interstellar travel technique presently available, since it uses known physics and known technology that is being developed for other purposes. Lightsails are being flight-tested for interplanetary travel, solar-pumped lasers and high power solar-electric sources for electrically powered lasers have been demonstrated for space power applications, and methods for combining many laser beams into one clean coherent beam and pointing that beam accurately have been demonstrated for space weapons purposes. Interstellar beamed laser propulsion systems will be larger than anything presently considered, but there is no new physics involved. Beamed power laser propulsion is energy inefficient, since most of the energy in the incident laser beam remains in the reflected beam and is not transferred to the vehicle. Better methods for accomplishing interstellar transport need to be found.

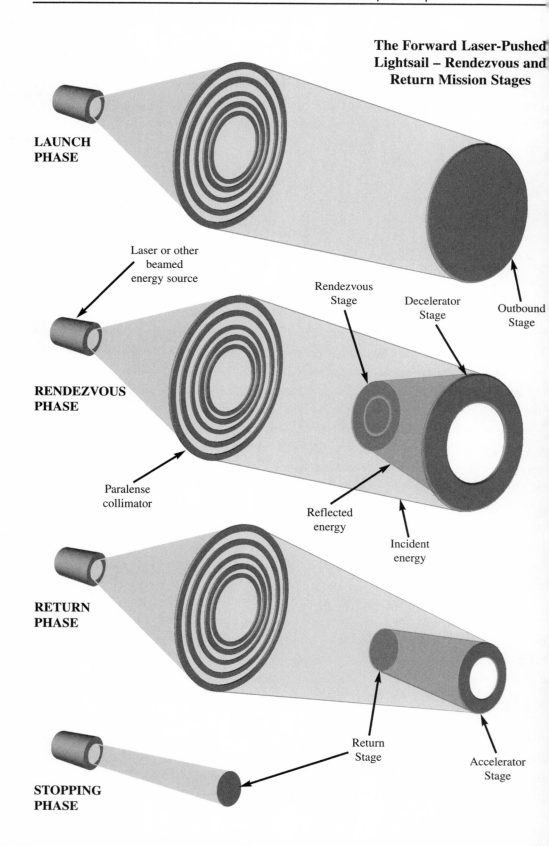

The Forward Laser-Pushed Lightsail – Rendezvous and Return Mission Stages

LAUNCH PHASE

Laser or other beamed energy source

Rendezvous Stage

Decelerator Stage

Outbound Stage

RENDEZVOUS PHASE

Paralense collimator

Reflected energy

Incident energy

RETURN PHASE

Return Stage

Accelerator Stage

STOPPING PHASE

The Forward laser-pushed lightsail (opposite page) is 1,000 kilometers wide, and would be propelled by light beams from powerful orbiting lasers powered by sunlight. Achieving a velocity of half the speed of light in 1.6 years, the starship would reach Epsilon Eridani in 21.6 years. Near the star, the inner two portions of the sail, containing the crew, would separate from the outer ring sail, be decelerated at Epsilon Eridani, and used as a solar-sail spacecraft to explore the area. The central portion would be used to return the crew back to Earth.

Research Topics Needing Further Study

For those interested in contributing to future studies of the problem of interstellar transport, these are some of the problems that need further study:

(1) Accurate calculation of the interaction of interstellar gas and dust on various materials (plastics, metals, plasmas, antimatter) and structures (solids, thin film sails, and gas/plasma/magnetic/multilayer shields) on near relativistic spacecraft at various speeds.

(2) Theoretical modeling and experimental measurement of the complete electromagnetic properties (reflectivity, absorptivity, transmissivity, and emissivity) of self supported, not oxidized, thin films and perforated meshes.

(3) Engineering limitations on pointing and tracking of beamed power systems over interstellar distances.

(4) Low mass apertures for high resolution imaging sensors on microrobot probes from microwaves to ultraviolet (replace lenses and dishes with imaging systems based on thinned phased arrays, speckle interferometry, holography, or synthetic aperture radar).

(5) Novel antennas for beamed power systems (how well can Fresnel, yagi or phased conform array antennas replace parabolic dishes?).

(6) Low mass beamed power receiving antennas for microrobot probes.

(7) High directivity, high efficiency, low mass, low cost power transmitting antennas.

(8) Theoretical and experimental limitations on the divergence of particle beams over interplanetary and interstellar distanced.

(9) Low capital cost prime electrical power in space.

(10) Solar-light-pumped prime sources of coherent microwave and optical power.

(11) Theoretical and observational limits to planetary imaging from the solar system. (If you can read the headlines on the *Alpha Centauri Daily Times* from an Earth-based telescope, do you really need to travel there?).

(12) Minimum science goals of the first flyby robotic probe.

(13) Shrink size of components on flyby spacecraft consistent with minimum science goals.

(14) Laboratory and orbital demonstrations of critical interstellar transport technologies (antimatter, magnetic sail, laser sail, microwave sail, fusion, etc.)

(15) Multiple miniprobe approach to interstellar missions using "shotgun" and/or "leader-follower" tactics.

(16) Scoops for interstellar ramjets.

(17) New energy sources.

(18) New non-rocket propulsion concepts.

10. Summary

It is difficult to build a starship, but it is not impossible. There are not one, but many future technologies under intensive development that can be used to produce a robotic starship that can reach the stars. All the feasible ideas seem to involve large and expensive systems in space. Better ideas for building an inexpensive interstellar transport system are needed. But until that time, it is helpful to know that there exist more than one method for making robotic starships suitable for travel over interstellar distances within a human lifetime.

References

1. R.L. Forward. A Programme for Interstellar Exploration. JBIS, 29 (10): 611-632, 1976.
2. R.P Feynman, R.B. Leighton and M.L. Sands. The Feynman Lectures on Physics, Reading Mass; Addison-Wesley Publishing Co., Inc. (Vol. 1, Chapters 15 & 16),1963.
3. J. Ackeret. On the Theory of Rockets. JBIS,6 (3): 116-123,1947.
4. R.L. Forward, E.F. Mallove, Z. Paprotny, J. Lehman and J. Prytz. Interstellar Travel and Communication Bibliography. JBIS, 33 (6): 201-248; 1982 Update, 36 (7): 311-329;1984 Update, 37 (11): 501-512;1985 Update, 38 (3): 127-136; 1986 Update, 40 (8): 353-364.
5. R.L. Forward. Feasibility of Interstellar Travel: A Review. JBIS, 39 (9): 379-384, 1986.
6. K.T. Nock. TAU – A Mission to a Thousand Astronomical Units. Paper 871049 presented at the 19th AIAAIDGLR / JSASS International Electric Propulsion Conference, Colorado Springs, Colorado, 11-13 May, 1987.
7. A.R. Martin and A. Bond. Nuclear Pulse Propulsion: A Historical Review of an Advanced Propulsion Concept. JBIS, 32 (8): 283-310, 1979.
8. J.C. Solem. Medusa: Nuclear Explosive Propulsion for Interplanetary Travel. JBIS, 46 (1): 21-26,1993.
9. F.J. Dyson. Interstellar Transport. Physics Today, October: 41-45, 1968.
10. R.L. Forward and J. Davis. Mirror Matter: Pioneering Antimatter Physics. New York: John Wiley & Sons, Inc. 1988.
11. B.W. Augenstein, B.E. Bonner, F.E. Mills and M.M. Nieto, eds. Antiproton Science and Technology. Singapore: World Scientific, 1988.

12. L.R. Shepherd. Interstellar Flight. JBIS, 11(4): 149-167, 1952.
13. B.N. Cassenti. Optimization of Relativistic Antimatter Rockets. JBIS, 37(11): 483-490, 1984.
14. A.R. Martin, ed. Special Issue on Antimatter Propulsion. JBIS, 35 (8): 283- 424, 1982.
15. R.W. Bussard. Galactic Matter and Interstellar Flight. Astronautica Acta, 6 (4): 179-194, 1960.
16. R.L. Forward. Roundtrip Interstellar Travel Using Laser-Pushed Lightsails. Journal of Spacecraft and Rockets, 21(2): 187-195. 1984.
17. R.M. Zubrin and D.G. Andrews. Magnetic Sails and Interplanetary Travel. Journal of Spacecraft and Rockets, 28 (2): 197-203, 1991.
18. R.M. Zubrin. Magnetic Sails, Unknown Nearby Stars, and Interstellar Missions. Proceedings Conference on Practical Robotic Interstellar Flight: Are We Ready?, New York, 19 August - 1 September, 1994.
19. D. Andrews. Cost Considerations for Interstellar Flight. Proceedings Conference on Practical Robotic Interstellar Flight: Are We Ready?, New York, 19 August - 1 September, 1994.
20. G. Landis. Laser-powered Interstellar Probe. Proceedings Conference on Practical Robotic Interstellar Flight: Are We Ready?, New York, 19 August - 1 September, 1994.

The Ultimate Exploration:
A Review of Propulsion Concepts for Interstellar Flight

Geoffrey A. Landis
NASA John Glenn Research Center

Abstract

Human interstellar flight is a problem with significant challenges. Even "slow" travel requires speeds of thousands of kilometers per second, orders of magnitude greater speed than any existing or planned space probe. For an interstellar flight, propulsion is the critical issue. The required ΔV is so high that any conventional propulsion system in development today is inadequate, and the question of propulsion technology dominates all other technology considerations. Fusion rockets and laser-pushed sails are two of the technologies which are most likely to be practical for interstellar flight.

1. Background

Introduction

Human interstellar flight is a problem with significant challenges. The possible approaches to an interstellar journey range from "slow" travel, at 1% of the speed of light or less, where journeys will require hundreds or thousands of years, or "fast" travel, at 10% of light speed or higher speed, where journey durations will be on the order of a decade. Even "slow" travel requires speeds of thousands of kilometers per second, orders of magnitude greater speed than any existing or planned space probe. "Slow" travel will require either a many-generation trip, a technology to suspend or slow-down human metabolism, or a way to send the humans in a compact form, such as in a robotic form, or in bodies grown on-site, perhaps as clones of the original humans. This requires considerable advances in biology, nanotechnology, or robotics. "Fast" travel allows humans to travel in unaltered form, but puts considerable constraints on the propulsion system.

For an interstellar flight, propulsion is the critical issue. The required velocity is so high that any conventional propulsion system in development today is inadequate, and the question of propulsion technology dominates all other technology considerations.

How far are the stars?

Today, when every day's television viewing shows science fiction featuring star ships traveling from star to star in a day or two, it is important to start with a reminder of how distant the stars really are. In a project (Apollo) that was arguably the most complicated human endeavor in history, the farthest distance humans have ever traveled is 400,000 kilometers from Earth – one one-hundred millionth of the distance to the nearest star. The Voyager probe, the fastest object ever launched from Earth, is now 12 billion kilometers from the sun and moving at 17.5 km/sec. In units more appropriate to discussions of interstellar travel, it has reached a distance of 0.001 light years from the Earth, moving at 0.006% of the speed of light. If ice-age humans had launched a star ship to Alpha Centauri at the speed of Voyager, at the end of the last ice age 11,000 years ago… it would be less than one-sixth of the way through its journey. Voyager will reach the distance of Alpha Centauri in 70,000 years. (It should also be noted that Voyager is not, in fact, traveling in the direction of any nearby stars.)

Another way to see why interstellar travel is difficult is to examine the rocket equation. The mass ratio of a rocket is:

$$\frac{m_i}{m_f} = e^{\left[\Delta V / V_e\right]} \tag{1}$$

where m_i/m_f is the ratio of the mass of the fully-fueled vehicle at launch to the mass of the empty vehicle at the end of thrust, ΔV is the change in velocity, and V_e is the exhaust velocity of the rocket. The best chemical rockets achieve an exhaust velocity of about 5 kilometers per second, so to achieve a velocity of only one tenth of one percent of the speed of light (fast enough to reach the nearest star in a little over four millennia) the mass of fuel required would be about ten to the 26th times the mass of the spacecraft!

Obviously, chemical rockets are a completely inadequate technology for interstellar flight.

2. Evolving Concepts

Nuclear Rockets

The most straightforward concept to go beyond the adverse mass ratios of chemical rockets is to develop a nuclear rocket. Since nuclear reactions have far higher energy per unit mass than chemical energy, the exhaust velocity V_e is potentially much higher. In fact, the theoretical exhaust velocity is proportional to the square root of the energy per unit mass of the fuel:

$$V_e = \sqrt{2E/m} \tag{2}$$

(If preferred, the exhaust velocity can be expressed as a specific impulse, Isp, by the relation $I_{sp} = V_e/g$, where g is the gravitational acceleration at the surface of the Earth.)

A nuclear fission reaction has an energy considerably higher than that of any chemical reaction, and yields theoretical exhaust velocities that begin to make interstellar flight thinkable. Nuclear fusion reactions, such as the deuterium-tritium reaction, (D + T —> 4He) have theoretical exhaust velocities that are considerably better yet. However, it is important to note that the actual exhaust velocity achievable is limited by the temperature

$$V_e = \sqrt{3kT/m_{atomic}} \qquad\qquad (3)$$

where m_{atomic} is the molecular weight of the exhaust. In a conventional rocket even with hydrogen, the lowest molecular weight exhaust possible, thermal limitations of materials mean that a nuclear-fueled rocket that uses heated gas cannot achieve an exhaust velocity above about fifteen kilometers per second, only slightly higher than that of the hydrogen / oxygen rocket. This is why many proposals for nuclear rockets for interstellar flight use systems where the nuclear reaction occurs outside the rocket, and hence are less limited by the temperature.

Orion

The first concept for interstellar flight that could be considered feasible in that it seriously addressed the mass-ratio problem, was the *Orion* nuclear bomb powered spacecraft proposed by Stanislaw Ulam, Theodore Taylor, and others. The concept is that a nuclear bomb is exploded behind the spacecraft. The shock wave of the explosion impinges on a "pusher plate," which transfers the momentum to the vehicle (see figure 1). These explosions are then repeated as needed until the vehicle reaches the required velocity.

This spacecraft was analyzed as a concept for interstellar flight by Freeman Dyson in an article in *Physics Today* in 1968. He proposed upgrading the atomic bomb powered spaceship of Ulam and Taylor to a fusion-bomb powered spaceship, and calculated [Dyson, 1968] that such a spaceship could reach a velocity somewhere between 1,000 kilometers per second and 10,000 kilometers per second, depending on his assumptions, and thus reach the nearest star over a flight time between a century and a thousand years. While this is quite a long time by any human standards, it is a possible propulsion system that does not violate the laws of physics.

Thirty-five years after Dyson's article in *Physics Today,* variants of the Orion continue to be the only concept for launch of large payloads on interstellar trajectories that does not require inventing new technology [Schmidt *et al.* 2001].

Figure 1 – *Orion* concept for an interstellar spacecraft
Courtesy of NASA

Daedalus

Inspired by the concepts of nuclear pulse propulsion developed by Orion, members of the British Interplanetary Society developed a more reasonable concept for a nuclear pulse propulsion interstellar probe. From 1973 through 1978, a thirteen-member team worked on the design of "Project Daedalus," a fusion-powered interstellar probe designed for a fly-by mission across a 5.9 light year distance to Barnard's Star in fifty years [Martin 1978].

The Daedalus concept proposed that a one-gram pellet of a frozen mixture of deuterium and helium-three could be compressed and heated by high-energy electron beams, and induced to undergo a fusion reaction (D + 3He —> 4He + p), creating a fusion "micro explosion." Although the deuterium / tritium fusion reaction is more commonly proposed, since it is easier to initiate, the Daedalus propulsion team chose the deuterium / helium-three reaction because the primary reaction yields protons and alpha particles, which can be confined and directed by a magnetic nozzle, and not neutrons, which are unaffected by magnetic fields. Fifty billion such pellets would be exploded behind a two-stage vehicle.

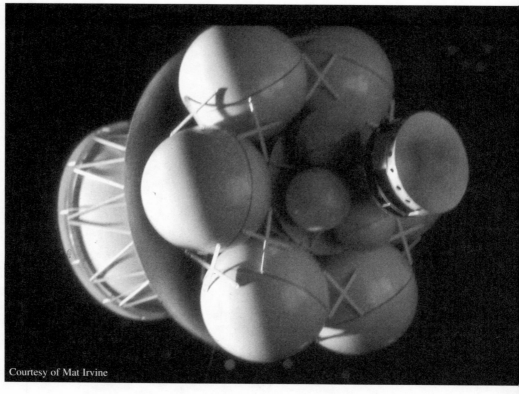

Courtesy of Mat Irvine

Figures 2a and 2b – The Daedalus interstellar probe concept [1978]

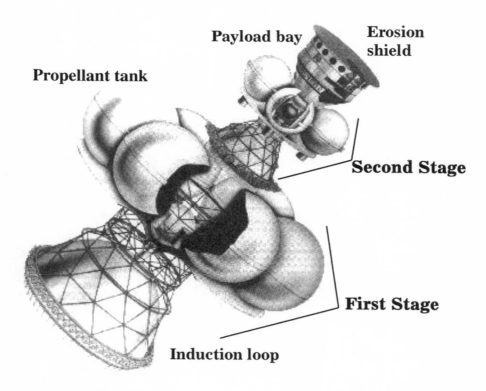

The difficulty of the helium-three propulsion reaction (in addition to the difficulty of initiating it) is that large quantities of helium-three are not easily available on the surface of the Earth. The Daedalus spacecraft required approximately thirty-thousand tons of Helium-three, and twenty thousand tons of deuterium, to propel a 450 ton payload to the desired cruise speed of 12% of the speed of light. To procure the helium-three, the Daedalus team proposed mining it from the atmosphere of Jupiter. Relying on a fusion reaction that has never been demonstrated, and requiring a fuel that requires mining missions to Jupiter, the Daedalus project was not a realistic endeavor for the near-term. Nevertheless, the Daedalus is a landmark in that the Project Daedalus team examined all the features of interstellar ship design, and the report remains the most complete study of interstellar ship design ever written.

Beamed Propulsion

Robert L. Forward revolutionized thinking about interstellar flight by developing new ideas that abandon the concept of using rockets. His analysis was that the mass ratio problem could be avoided by the simple expedient of leaving the actual power source for the spacecraft stationary. In a series of papers (as well as science fiction stories), he proposed and analyzed the concept of pushing spacecraft by beamed energy.

The concept of a solar sail was first proposed by Tsander in 1924. He noted that since a photon carries momentum, the reflection of a beam of light will provide a push, and hence a large thin reflective "sail" could be pushed by sunlight, resulting in a method of propulsion in space which uses no propellant (and hence can reach velocities which are not limited by reaction mass). The amount of force produced by sunlight is very small. It is set by the Einstein relation. For a sail which reflects a fraction R of the incident light (see figure 3 on next page) and absorbs a fraction A, the force on the sail, for the case of the sail perpendicular to the incident light, is

$$F = (2RP / c) + (AP / c)$$

where P is the incident power and c is the speed of light. The factor of 2 accounts for the fact that the reflected light transfers twice its momentum to the sail. For the case of a fully-reflecting sail, this reduces to $F = 2P/c$, or 6.7 Newtons per gigawatt.

While sunlight can be used to push a lightweight sail, the fact that sunlight decreases by the factor $1/r^2$ with distance from the sun puts a limit on the maximum velocity achievable in a trajectory directly outward from the sun. Matloff (2000) estimates this to be about 500 kilometers per second, or about 0.17% of the speed of light, for an example sail. This is far faster than existing propulsion systems, but inadequate for realistic interstellar missions.

The concept of a laser-pushed light sail is shown in very crude schematic in Figure 3. The idea was first proposed by Forward [1962] and Marx [1966], but the practical implementations of the concept were done by Forward [1984].

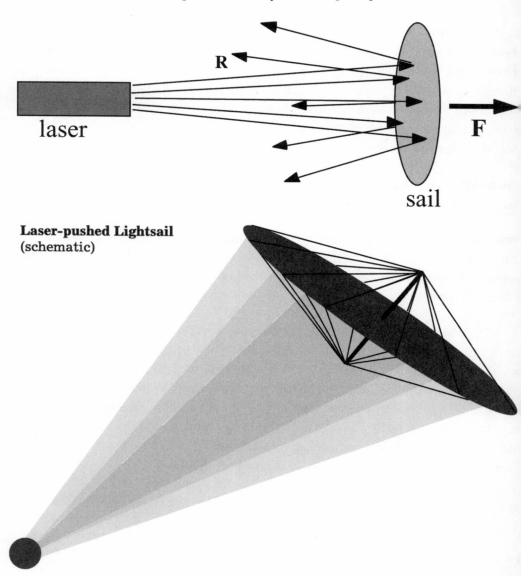

Figure 3 – Laser-pushed lightsail [schematic]

Forward proposed using a laser instead of sunlight to push a light sail. He provided background calculations on how a laser-pushed sail would work to science fiction writers Larry Niven and Jerry Pournelle for use in a science fiction novel [Niven and Pournelle 1974]. He then developed the concept in more detail, first in a science fiction novel [Forward 1982], and then publishing detailed calculations in a landmark paper in the *Journal of Spacecraft and Rockets* [Forward 1984]. As shown in table 1, Forward proposed three missions, using progressively more speculative technology. Even the smallest of the probes, the "flyby" mission, required a lens 1,000 kilometers in diameter to focus the laser, and required 65 GW of laser power.

The second, more advanced mission, proposed stopping at the target star. Since a laser is able only to push on a sail, stopping the probe at the target star is difficult. Forward proposed that the light sail separates into two parts, and the larger element reflects the incident light onto a smaller segment of the sail. In this way, the reflected light can stop the smaller sail. This technique has the difficulty that the laser must be focused to a small point at the distance of the target star, and a further difficulty that the sail surface quality must be extremely good in order to avoid diffusing the beam.

The third mission proposed using this technique not only to stop at the target star, but also to reflect the light to make a return trip.

Table 1 – Summary of parameters of three interstellar light sail missions proposed by Forward

Mission Stage	Laser Power (GW)	Vehicle Mass (tons)	Accel. (gees)	Sail diam. (km)	Maximum Velocity (% of c)
1. Flyby	65	1	0.036	3.6	0.11 @ 0.17 ly
2. Rendezvous					
outbound	7,200	785	0.3	100.	0.21 @ 2.1 ly
decel.	26,000	71	0.2	30.	0.21 @ 4.3 ly
3. Manned					
outbound	75,000,000	78,500	0.3	1000.	0.50 @ 0.4 ly
decel.	17,000,000	7,850	0.3	320.	0.50 @ 10.4 ly
return	17,000,000	785	0.3	100.	0.50 @ 10.4 ly
decel.	430,000	785	0.3	100.	0.50 @ 0.4 ly

Following Forward's pioneering 1984 light sail paper, several detailed analyses went on to analyze and refine the small light sail concepts. Several engineering difficulties in the concept were pointed out by Andrews and Zubrin [1988] and by Landis [1989]. The work by Andrews and Zubrin introduced the concept of a magnetic sail which can be used as a brake, with the magnetic field of the sail dragging against the interstellar medium to slow the vehicle down at the target star. They concluded that this might be a more practical method of stopping than the sail reflector.

Considerable analysis of both the laser and solar pushed interstellar missions was also done by Mallove and Matloff [1989] and by Matloff [2000].

Landis analyzed several variants on the Forward "Flyby" probe, concluding that the maximum operating temperature of the sail was a critical parameter in the performance. He concluded that shifting from an aluminum sail to Niobium would allow the probe to reach speed much faster, and hence allow a smaller probe [Landis 1995]. A later paper proposed using refractory dielectric materials for the sail [Landis 2000A]. Table 2 shows some of the evolving concepts. The last concept, by Jordin Kare [2001], uses an intense laser to accelerate tiny sails at ten million times the acceleration of gravity. These micro-sails are not individual probes, but rather are used as a stream of particles which then impact the probe. The use of a laser to accelerate microsails thus makes practical a pellet-pushed concept first proposed by Singer [1980].

Table 2 – Some evolving concepts for "small" interstellar light sail probes

Forward thin aluminum sail 1985   Power: 65 GW, acceleration: .036g (10% of lightspeed in 2.5 years)
Landis Niobium sail 1995   Power 107 GW, acceleration 0.82 g (10% of lightspeed in .11 years)
Landis dielectric sail 2000   448 MW, acceleration 43.4 g (10% of the speed of lightspeed 8.5 days)

Kare lasersail-pushed magnetic sail 2001
    25 GW, 10 million g (10% of the speed of light in 4 seconds)
    (microsails-- 26 centimeter diameter, 16 milligrams each)

In 1985 Forward analyzed the alternative possibility of using microwaves, rather than lasers, to push a lightweight probe [Forward 1985]. This lightweight microwave-pushed probe concept was analyzed in further detail by Landis [2000B].

Landis [1989 and 2001] proposed the use of a particle beam instead of a laser beam to push the sail. The particle beam impinges on a magnetic field (a "magnetic sail"), as proposed by Andrews and Zubrin [1988]. The magnetic sail has the advantage of being primarily empty space; the magnetic field is created by a loop of superconducting wire. The advantage of the particle beam is that it is not subjected to diffraction, and hence does not require the 100 to 1,000 kilometer lens needed by the laser systems.

Conclusions

This overview has given but a brief (and by no means exhaustive) summary of interstellar propulsion concepts.

In 1968, when Freeman Dyson published his article on adapting the Orion nuclear-pulse rocket concept to interstellar flight, he quoted a well-known physics professor as saying "All this stuff about traveling around the universe in space suits – except for local exploration, which I have not discussed – belongs back where it came from on the cereal box." [Dyson, 1968]. In the intervening 37 years, we've come a long way toward turning cereal box dreams into physics reality, but I still believe that there are still a lot of concepts yet to be invented, which may, someday, turn interstellar flight into engineering reality.

References

D.G. Andrews and R.M. Zubrin (1988), "Magnetic Sails and Interstellar Travel," IAF Paper IAF-88-5533, 39th IAF Congress, Bangalore India, October.

F.J. Dyson (1968), "Interstellar Transport," *Physics Today,* October, pp. 41-45.

R.L. Forward (1962), "Pluto, Gateway to the Stars," *Missiles and Rockets, Vol. 10,* p. 26 ff. (April); reprinted *Science Digest 52*, pp. 70-75 (August 1962).

R.L. Forward (1982), "Rocheworld (part 1)," *Analog Science Fact / Science Fiction,* December.

R.L. Forward (1984), "Roundtrip Interstellar Travel Using Laser-Pushed Lightsails," *Journal of Spacecraft and Rockets, Vol. 21,* Mar-April, pp. 187-195.

J.T. Kare (2001), "High-Acceleration Micro-Scale Laser Sails for Interstellar Propulsion." Report of NIAC Phase 1 project, October (NASA Institute for Advanced Concepts).

G. Landis (1989), "Optics and Materials Considerations for a Laser-propelled Light sail," Paper IAA-89-664, 40th IAF Congress, Torremolinos Spain, Oct. 7-13.

G. Landis (1995), "Small Laser-propelled Interstellar Probe," Paper IAA-95-4.1.1.02, *46th IAF Congress,* Oslo Norway, October; published *Journal of the British Interplanetary Society, Vol. 50,* No. 4, pp. 149-154 (1997).

G. Landis (2000A), "Dielectric Films for Solar- and Laser-pushed Lightsails," *AIP Conference Proceedings Volume 504*, pp. 989-992; Space Technology and Applications International Forum (STAIF-2000), Jan. 30 - Feb. 3, Albuquerque NM.

G. Landis (2000B), "Microwave Pushed Interstellar Sail: Starwisp Revisited," paper AIAA-2000-3337; 36th Joint Propulsion Conference, Huntsville AL, July 17-19.

G. Landis (2001), "Interstellar Flight by Particle Beam," *AIP Conference Proceedings Volume 552*, pp. 393-396; STAIF Conference on Innovative Transportation Systems for Exploration of the Solar System and Beyond, Feb. 11-15, Albuquerque NM.

E. Mallove and G. Matloff (1989), *The Starflight Handbook,* Wiley Science Editions; John Wiley and Sons.

G. Matloff (2000) *Deep Space Probes,* Springer Verlag / Praxis publishing,

A.R. Martin, ed. (1978), *Project Daedalus – Final Report,* special issue of the *Journal of the British Interplanetary Society*, January.

G. Marx (1966), "Interstellar Vehicle Propelled by Laser Beam," *Nature, Vol. 211,* July, pp. 22-23.

L. Niven and J. Pournelle (1974), *The Mote in God's Eye,* Pocket Books.

G.R. Schmidt, J.A. Bonometti, and C.A. Irvine (2001), "Project Orion and Future Prospects for Nuclear Pulse Propulsion," paper AIAA-2000-3865, *J. Propulsion and Power, Vol. 18*, No. 3, May-June 2002, pp. 497-504.

C.E. Singer (1980), "Interstellar Propulsion Using a Pellet Stream for Momentum Transfer," *J. British Interplanetary Soc., Vol. 33*, Mar., pp. 107-115.

Colonizing Other Worlds

Joe Haldeman

Before going about the business of how humankind might go about colonizing other worlds, it's relevant to ask why. Why even consider such a far-fetched, possibly absurd, project in the first place?

There are two unrelated sets of reasons for colonizing other planets – one set would be because we want to, and the other set, because we have to.

The "have to" playbook is of course the one that fiction and drama prefer – the killer asteroid, the nuclear winter, the ultimate virus. Books and movies on the basic theme predate the scientific models, such as discovery of the dinosaur-destroying Chicxulub meteorite. If you're over fifty, you might remember being terrified by the movie *When Worlds Collide* (if you're under forty, you would more likely be amused by the cheesy special effects). It had a rogue star come hurtling out of nowhere straight toward the Earth. The star fortunately had an Earth-like planet hurtling obediently right alongside it, so it was relatively easy for some determined American scientists to cobble together a space ship fleet and send a herd of stalwart breeding stock over to the new world.

Of course the probability of that cosmic collision and coincidence is exactly zero, but we do have more interesting, and more likely, doomsday scenarios nowadays. We can evidently expect a catastrophic collision like the Chicxulub asteroid or comet every few million years; applying everyday actuarial mathematics to that number gives us the cheerful news that, in any given year, an average person is more likely to die from a killer meteorite than from a plane crash. Maybe it would be prudent to have an escape hatch in the form of an orbiting generation ship.

There's also the threat of all-out nuclear war, of course, though many experts say it would take more than the tens of thousands of extant nuclear weapons to actually get rid of everybody. More likely, I think, would be an unexpected side effect of a new weapon that hadn't been adequately tested, like a biological agent with vectors that reach everyone in the world before the first symptom presents itself. (It's worth remembering that some of the scientists on the Manhattan Project wondered whether the atom bomb might set off a universal chain reaction, destroying the Earth, but they weighed that small probability against the threat of an Axis victory, and pushed the button anyhow.)

A favorite science fiction theme in the sixties and seventies was a future Earth made uninhabitable by runaway pollution. Now that we're in the future, we see how silly that was. We may eventually wind up with a world so unpleasant you wouldn't want to live there, but that belongs to the other set of scenarios.

Then there's the nanotechnology "gray goo" nightmare, where the tiny machines turn Sorcerer's Apprentice and reduce all Earth to a uniform product. Nanotechnology experts claim that couldn't happen; that safeguards would be built in to keep nanomachines from reproducing themselves *ad infinitum*. They're probably right; the whole world didn't explode after the Trinity test, either.

(Unfortunately, if "gray goo" does happen, it won't be slow. There won't be time to build a Buick, let alone a generation ship. Maybe there will be humans in orbit or on other planets in the solar system, though, who can escape the little devils.)

My own favorite doomsday scenario – I wrote a book, *Forever Peace*, about it – would be the discovery of a process or principle that would enable anyone to destroy the world, just by wanting to do it. I personally would like to be far away before it became common knowledge. If the oceans were suddenly turned to gasoline, how long do you think it would be before someone lit a match, just to be the one who did it?

Even if the doomsday machine were as hard to acquire and use as a nuclear weapon is today, it would only be a matter of time – perhaps centuries or millennia, but eventually – before a suicidal megalomaniac leader got a hold of one and decided to take everyone with him. The inevitability of that situation might give a sane part of humanity both time and motivation to build the ship and light out for the stars.

As dramatic as these emergencies and insanities are, I think it's actually more interesting to consider situations less dire, where we send a ship to the stars not because we have no choice, but because it looks like an interesting and worthwhile enterprise.

Back in 1969, Freeman Dyson, as an "entertainment between semesters," did some back-of-the-envelope calculations approximating the cost of a generation ship. This was Theodore Taylor's design, the *Orion* project, which used nuclear bombs to propel a large space ship up to a speed around 10,000 km/sec. This is about one parsec per century, so there would be dozens of target stars within a few centuries' flight. (There are 74 within five parsecs.)

He came up with a price of 10^{11} dollars, for a payload of 10,000 tons. That was a large fraction of the Gross National Product in 1969. In the long run, though, that didn't bother him. "If we are thinking on a time scale of centuries," he said, "our GNP is far from being a fixed quantity. Presumably the human race will either destroy itself or continue its economic growth at something like the present rate of 4% per year. If we destroy ourselves, space ships are not going to be of interest to the survivors for quite a long time. If we continue our 4% growth rate we shall have a GNP a thousand times its present size in about 200 years. When the GNP is multiplied by a thousand, the building of a ship for 10^{11} will seem like building a ship for 10^8 today. We built a fleet of Saturn V's, which cost about 10^8 each. It may have been foolish, but we did it anyhow. On this basis, I predict that about 200 years from now, barring a catastrophe, the first interstellar voyages will begin."

In fact, the Gross Domestic Product (which is the GNP minus income from abroad), measured in 1996 dollars, increased from $2.38 trillion to $8.85 trillion in the past forty years, which represents slightly less than 3.5% compound interest, so it might take a little longer than 200 years for a star ship to be a reasonable enterprise. But the argument is the same, sarcasm and optimism in about equal measures.

Given the physical possibility of such a trip, why would it be desirable? Historically, people have left, or fled, because conditions abroad were better than at home – a potato famine at home or El Dorado abroad – but those conditions don't make economic sense here. Instead of spending the equivalent of 10^8 to send a thousand people on a trip, you could give each of them $100,000 worth of potatoes or gold. Neither would it make much sense for a nation to do the interstellar equivalent of emptying out the prisons to colonize Australia, at a hundred grand per felon.

It's safer to deal in abstractions. A lot of people, especially Americans, will want to go to a new place just because it's there and underpopulated.

Frederick Jackson Turner wrote in 1894 that the form and spirit of American democracy were directly due to the presence of the frontier – free land, encouragement of individuality, ingenuity, and resourcefulness. To what extent might that apply to space colonization?

Of course the first big difference is the scale of the enterprise. In frontier days, a couple could hitch a mule to a wagon, throw in an axe and a shovel and a rifle, and walk west until they found a place to settle. It's hard to imagine any propulsion technology that would allow that sort of casual diaspora to the stars, though truly cheap fusion might make possible a kind of land rush out to the Moon, Mars, the asteroids, and the satellites of Jupiter and Saturn.

I was surprised to calculate that if that couple could make twenty miles a day, and took Sundays off, in forty years they could walk the distance to the Moon – a quarter of a million miles in the time that Moses and his followers wandered around the Middle East.

Trying to apply that human scale to the rest of the Solar System, though, you wind ɔ with periods of time that are long even by generation ship standards. If there were a ›ad to Mars, it would take 8,000 years to walk it.

That's not completely irrelevant. If someone started walking to Mars in 1850, in ɔout fifty years they would be passed by people in cars. A few years after that, rplanes. In another half century, actual space ships, reaching escape velocity.

Who's to say that the same thing wouldn't happen to a generation ship? You start ɪt on a centuries-long project, and before you've gone a tenth of the way, someone ɪsses you with a faster, cheaper way to travel. By the time you get to Alpha Centauri, ːople are commuting from Earth by wormhole, and the virgin wilderness is all suburbs ɪd shopping malls. The possibility of that happening could delay a generation ship art-up indefinitely.

Suppose none of that does happen, though, and the simplest model comes to pass: e've identified an Earth-like planet fairly nearby, and have the wherewithal and ɔtivation to send a generation ship off in its direction. What would you need in order set up shop?

An assumption behind most schemes is that the main expense, by far, of getting ɔlonists to an extrasolar planet would be transportation, so you would spare no expense the details of assembling the colonists and their gear. But there's one big constraint: r any propulsion system we're familiar with, you want to minimize the size of the ɪyload.

Does this mean to minimize the number of colonists? If you postulate a completely ficient closed ecology, everything recycled, then the mass associated with life support constant for a given population, no matter how long the voyage, and probably small ›mpared to the mass of the entire vehicle. (Of course, there's no such thing as 100% ficiency. With inevitable loss of water and atmosphere through leakage, you'd have to ɪrry spare resources or mandate negative population growth en route. The energy to ɪndle photosynthesis also requires expenditure of mass, though maybe as small as ːcreed by $E = mc^2$.)

Most stories about generation ships assume that a population the size of a small ɪwn will be inside the star ship – thousands of people with hundreds of necessary skills, illing and able to pass on their knowledge to the next generation, and the next. I've ritten that one myself.

But it bears examination.   Suppose you postulate, reasonably, that it will b centuries in the future before the human race is able to take on a project so expensiv and quixotic as a generation ship. What will we have learned by then?

It's not absurd to think that we may have learned a lot about learning – that it migh be possible to teach anybody anything, whether they are initially interested in it or no It could be that we would also know a lot more about creating people to order, an instead of hoping to educate a few generations of living breathing sanitary engineer say, we might just ship the proper egg and sperm (or blastocycst-in-waiting) along wit a sanitary-engineer expert system, to wait until a few years before landing, create properly motivated girl or boy, and teach it exactly what it hungers to learn about bein a sanitary engineer.

One problem with an Earth-like planet, where these kids can just charge out wit axes and mules, or robots, to forge a new civilization, is that if a planet can provide foo for us, we can be food for its natives. This was a common theme in exploration storie in my father's time – alien planets were bound to be full of carnivores who alway curiously, had a taste for human flesh. (Most of the planets, though, like Edgar Ric Burroughs' Barsoom, didn't have any herbivores for the carnivores to eat. So it's n wonder they weren't picky – they were starving!)

We'd probably be more likely to face problems at the other end of the size scale microscopic pathogens who find our proteins just as congenial as the native ones. (Th idea predates Barsoom, of course – it was the common cold that defeated the implacabl Martian invaders in Wells' *War of the Worlds*.) Of course we will probably know enoug about the human genome by then to whip up a vaccine before the first human opens th airlock.

Of course the colonists would have much more sophisticated hardware than ax and mules. A big problem with trying to guess what tools a human colony would brin to a new world is that we seem currently to be on the brink of resolving two real central technical questions – either solving them absolutely or finding a limit to the usefulness. Those are nanotechnology and life extension.

Nanotechnology has been demonstrated, of course, in processes as mundane a cooking surfaces and self-cleaning laundry. But we're still orders of magnitude awa from what we actually need: Drexler-style self-replicating nanomachines.   Lif extension has yet to be demonstrated, least of all at the scale we need for centuries- c millennia-long voyages, and arguably it awaits nano- and picotechnology itself.

There's nothing wrong, in terms of a thought experiment, with just waving a wand and saying "Make it so." Immortality, or near-immortality, could replace the generation ship with one that has a single crew, of exquisitely trained and presumably rock-stable people. NASA is evidently worried about the possibility of murder on a mere three-year Mars trip. After three centuries you might have one survivor, playing Lego with the bones of his crew mates. Or you might have a hundred people placidly playing their hundred-millionth rubbers of bridge. I leave it to you which nightmare is worse.

A midway scenario would involve something like hibernation, where you could pick the crew of the star ship and train them on Earth, and then have them sleep away the centuries in transit. We don't yet know how to put people into suspended animation, but it's a good bet that we'll be able to do that before we can bestow immortality or thaw out the dead.

If humans were long-lived enough, we might commit ourselves to projects that would take thousands of years of start-up effort, in which case we could send self-replicating probes out to hundreds of Earth-like worlds, and have them modify the places to human standards, and set up bases awaiting human arrival. Let the machines do the dirty work.

By then, though, it may be that the self-replicating machines would be as human as we are. So why send old-fashioned flesh humans at all?

This could lead to a kind of "virtual colonization," with our surrogates sending back information about what it's like living on their worlds. If the information were precise and in the right form, a human could plug into it and experience an illusion of being here that would be indistinguishable from the input our senses give us of being here – which is also a systematic illusion, reality filtered through the limitations of our senses and time scale. Of course, we couldn't "tele-operate" our surrogates from light years away, but given an arbitrarily large amount of information about the environment, it would be child's play – a futuristic Game Boy – to generate realistic feedback situations so that we could physically explore the new world, discovering second-hand the novelties that our surrogates first encountered.

Here's a science-fictional twist: having delivered that information, our surrogates might have the power to change their form so that they resemble native animals, or even plants or mineral formations, and patiently observe, waiting for something to happen. Maybe some other race did this on Earth millions of years ago, and elephants or coral reefs or Mount Rushmore have been watching us evolve. Maybe they sent a warning signal off to their creators when humans first left this planet, and we'll soon have visitors.

Frank R. Tipler has extrapolated the possibilities of exploration via nano-machine with a vengeance, in his "Deus ex Silico" model.

He defines the human soul as a tremendously complex computer program run on the wetware human brain. In his near-future scenario, each soul, for the purpose of transportation, can be uploaded into a tiny quantum computer, and by this means, thousands of humans can be fit into a container that masses only a few grams, and needs no heavy and inelegant life support system.

A tiny payload only needs a tiny space ship, in Tipler's model. Waving the wand again, he gets matter-antimatter annihilation, which can provide constant acceleration, moving a one-kilogram mini-ship to Alpha Centauri in only five years.

Once it's there, of course, it would be silly to download the personalities into squishy mortal human bodies. The souls are what make them human.

"Human uploads have such a natural advantage over present-day people in the environment of space," Tipler says, "it's exceedingly unlikely flesh-and-blood beings will ever engage in interstellar travel."

From there it's only one more wave of the wand to have the space ships be self-replicating themselves – each one makes two more space ships, with the same payload of thousands of souls. They take off for two more planets, and each does the same thing. In this manner, he claims, "the entire Milky Way will be explored and colonized in less than 1 million years. In another 10 million years, the Local Group of galaxies will be explored and colonized. In another 100 million years, the Virgo cluster of galaxies will be completely colonized. Earth will be entirely dismantled to provide the raw material for the expanding biosphere, long before its scheduled rendezvous with the expanding sun 5 billion years out. The beauty of the colonization process is that it's exponential, completely engulfing the universe with computer-borne human life 10 million trillion years from now."

Unless some other creature does it first.

Interstellar Travel: Why We Must Go†

Doug Beason
8515 Vina del Sol NE
Albuquerque, NM 87122

Humans have always dreamed of going to the stars. But aside from appealing to
the human spirit or the future need for *lebensraum*, little has been done to detail the
socio-political and cultural reasons for undertaking such an enormous venture, to show
why we *must* go.

With the exception of any science fictional faster-than-light travel, we've almost got
the technology to go to the stars; some argue that we've already got it.

But rarely does just having technology "push" a requirement. In fact, this is known
as a solution looking for a problem. Just because we have the technology for interstellar
travel does not mean we're going to go. For example, we've had the technology for
years to go back to the Moon – but since the end of the Apollo program, human presence
has been limited to a mere few hundred kilometers from the surface of the Earth.

More often than not, as the old saying goes, "necessity is the mother of invention."
Thus, mounting an interstellar expedition includes garnering popular and political support.

This article lays the groundwork for showing the overwhelming need for mankind
to embark on interstellar travel. In addition, it brings to light two enabling technologies
that can make the mission possible. It is these arguments, fueled by the exciting new
advances being made in our research laboratories, that could make this dream of
traveling *ad astra* come true.

*****

Over forty years ago President John F. Kennedy challenged the nation to launch the
most ambitious project ever undertaken by man: to land a man on the Moon within a
decade and bring him safely back. It was a grand vision, and history paints the era as a
galvanized country gearing up for the program and standing firmly behind the President.

Although it was true that many were swept up in the grandeur of Kennedy's vision,
the majority of the world was skeptical that this lofty goal could ever be achieved. Nay-
sayers in our own country bubbled out of the woodwork, giving reasons for the
President's motivations that ranged from forwarding political agendas to serving as a
"front" for America's infant intercontinental ballistic missile technology.

Some of the leading scientific minds of the time argued that there were too many insurmountable problems that stood in the way. A few of the objections raised that have since been proven untrue include arguing that because of technology limitations and orbital dynamic requirements it was impossible for spacecraft to rendezvous with other spacecraft; or that the Van Allen belt would subject the astronauts to lethal radiation; or because excess heat could not be dumped in the vacuum of space, the astronauts would be boiled alive in their spacecraft; and even that the lunar dust was too fine to support any weight and a landing craft would sink into the lunar surface.

Fortunately, common sense prevailed and the nation was not misled by every criticism brought against the program. Rather, a logical engineering path was laid out that identified the risks, and the Apollo program was executed with sufficient funding and popular support. (Contrast this to the current national missile defense program, that has popular support but insufficient funding.)

Now fast forward to July 20, 1989. On the steps of the Air and Space Museum President Bush gave a speech commemorating the 20th anniversary of the landing of the first men on the Moon. In that speech he outlined another exhilarating goal, the Space Exploration Initiative – a goal for humans to return to the Moon, this time to stay, and then to go to Mars *and beyond*. This marked the first time since John F. Kennedy speech that America was challenged to "go where no man has gone before."

For all intents and purposes, America had pulled out of exploring space at the end of the Apollo era. If properly instituted, Bush's new challenge could define renaissance for exploring first our solar system, then the universe.

The crux of the Space Exploration Initiative was to pursue the peaceful application of high technology. With a goal of first reaching Mars, it would have inspired and invigorated generations to come. And when combined with technical spin-offs, the increase in scientific knowledge would have reestablished national leadership in science and technology.

However, the Space Exploration Initiative met a quick death. After the end of the Cold War and especially immediately after the Gulf War, the politicians could have electrified the nation by undertaking such an adventure. Instead, the initiative diffused away, lacking popular and thus political support.

What was wrong with the plan? Was some technical flaw discovered that eventually led to its downfall?

It wasn't that it was technically infeasible. Some of the best scientific minds of the country had been assembled to hammer out the architecture that would eventually send humans to Mars and beyond.

What was missing was the logical foundation underpinning the program. As the Space Exploration Initiative would end up diverting national resources and thus compete for a large portion of the government's discretionary budget, in the years of "zero sum budgets" (for every new program, an old one dies), the case had not been adequately made. Instead, it was a return to "business as usual," and the opportunity to return to the Moon and on to Mars was lost.

But that was over a decade ago.

Today, economic and political conditions are in turmoil and there is much movement to attempt another redirection of national resources. With the bear market running its course and the war against terrorism focusing our defense infrastructure, just as other economic downturns and wars have provided enormous opportunity for growth, the next few years will be crucial to building momentum for a return to space.

But just as critics existed in the Apollo program 40 years ago, it's even a tougher sales job with today's instant world-wide communication and access to the mass media. As a result, the two largest criticisms need to be addressed up front: why bother going to the stars, and why now?

The obvious answer for interstellar travel is that the program would provide a long-term focus for not only our space efforts, but for the nation as a whole. It allows us to invest in our nation's scientific and technological base – traditionally the best product that America has to offer in this post-industrial age era. This will create new job opportunities and markets in the new economy. With the nation weary of riding an economic roller coaster, this allows for long-term, stable investments. Simply arguing to go back into space for the benefits of some future "Tang and Teflon" doesn't garner national support.

More directly, an interstellar program provides the chance for our nation to reorient our GNP from depending on both the military-defense establishment and the latest "killer app," in business' always-fluctuating quest for maximizing short-term profits. This is especially critical during these post-911 days, when America needs a long-term, stimulating national goal on which to base its economy.

An interstellar program has the opportunity to facilitate the commercialization of space and promote space-based industries, products, and services. Further, it would allow for advancing technological innovation.

Once the requirement for a long-range interstellar program is established – the "necessity" part of that old adage – then new technologies with terrestrial and commercial applications will abound. And as shown by previous large investments in major national undertakings (such as the Apollo program), this in turn will inspire a renewed interest in science and engineering.

The consequence of investing in a long-range program for interstellar travel is to radically increase our knowledge of the Universe. This will help us better understand the origin of the Universe, stars, planets, and perhaps life itself.

These are all good reasons for establishing a major initiative for interstellar travel. But the question remains, why should we do this now? Why can't we wait until there are better times, economically as well as socially? After all, it's pretty hard to justify an exploration program when there are plenty of people hurting right here at home.

Part of the reason has already been given: an interstellar program is a fundamental way to ensure that those better times will come. It's more than sending people to the stars. Rather than simply dumping money into the hands of bureaucrats, "experts" in redistributing wealth, it provides for a systematic way to rebuild our nation's infrastructure.

In the 1940's, after World War II, America invested in the future of its past adversaries, Germany and Japan, through such policies as the Marshall Plan and the MacArthur accords. Some, looking back, may say that since Germany and Japan are now our economic rivals, that this was not such a good idea. But there is no doubt of the success these programs have had in rebuilding those nations' infrastructure.

America is in a similar situation today. But this time we have a chance to invest in *ourselves*, to rebuild the very backbone that defines our nation's economic strength. With the demise of the internet economy, and the nation rebuilding both physically and psychologically, it's time to return to the basics: stable growth. The message is that long-term investment *works*.

And there are other compelling reasons.

An interstellar program is more than throwing money at exploration: it embodies the essence of a new social paradigm and it can therefore be constructed without limits. This is different from saying that "no end is in sight." Rather than having a closed goal of solely sending humans to the stars and ending the program there, a true interstellar program will use education as a basic priority to bring out the very best in our nation, and to hold the future open for our children. It is forward-looking and will focus chnologies to allow this social effort to succeed. (Contrast this with what happened to e Apollo program: once we got to the Moon, the public lost interest and the space ogram went downhill.)

Interstellar travel provides a focus to the entire space program. Just as Apollo icceeded in the 1960's, this can unify our future goals in space, which correspond to e types of missions that may be chosen: exploration, science, and colonization.

Interstellar exploration is the classic "flags and footprints" reason for going to the ars. It can start to answer that question of "are we alone in the universe?" while tisfying the basic human need to explore the new frontier.

As a prelude to mounting an interstellar program, testing equipment on the Moon ovides a logical first step. Since the Moon is only 240,000 miles away and has a irface area of 14.6 million square miles (roughly that of Africa), it provides a unique st bed for checking out equipment and procedures while being only three days away. iis insures that everything is fully tested in an environment "not too far from home." ie unusual success of the Apollo program was because it was based on the philosophy at *everything* should be simulated as accurately as possible ahead of time. Surprises ould not be tolerated. A good analogy is that when buying scuba equipment it's wise check out your tanks in a swimming pool before diving to the bottom of the ocean.

An interstellar science program will increase our knowledge of the Universe. dvances in stellar and planetary evolution, space science, cosmology, cosmic radiation, d interstellar field dynamics are just a few of the basic science fields that would benefit.

As a short example, an entire part of the electromagnetic spectrum (sub-millimeter) forever hidden from observation on Earth because of the ionosphere's plasma equency. Moving beyond the Earth's atmosphere would allow incredible opportunity r astronomical observation in this frequency region. As the exploration craft moves vay from Earth and beyond Jupiter, an even lower part of the spectrum (megahertz) ould be shielded from the interference from those planets' electromagnetic noise.

More directly, low-energy cosmic rays are prevented from being observed on Earth nd even in the solar system) because of the Earth's and Sun's magnetic field (the geomagnetic cutoff:" low-energy charged particles follow helical paths around the agnetic field lines and are eventually "funneled" into the poles. An absence of a agnetic field allows for a pure measurement of both the direction and energy of these rticles, giving insight into their origin.)

Finally, there are a plethora of traditional instruments that would benefit from being in the pristine environment of interstellar space. Worries about thermal stresses, gaseous impact ionization, debris, and obscuring multi- and hyper-spectral radiation would all be alleviated. And the opportunity would exist to find the answer to such basic puzzles as the interstellar gas density, the hydrogen contribution to the cosmological red-shift, and perhaps even quantum zero-point energy fluctuations.

Interstellar colonization is the ultimate goal for a journey to the stars. As noted in the beginning of this article, until a (for now) science fictional way is found to travel faster than-light, interstellar expeditions incur a lifelong commitment. By not tying humans to the relative frailty of our solar system – spanning wars to "epoch changing events" such a major asteroid impact – colonization insures the survival of the human species.

*****

Technical Considerations

Crew safety is the number one priority. It just wouldn't do to mount an interstellar expedition only to have the crew not survive. This dictum of "safety first" is just as true for going to the stars as it was for Apollo.

Back in the 1960's, the Apollo program had a specific goal of within a decade putting a man on the Moon and returning him safely. An interstellar program is much longer, with considerably more social and fiscal pressure, so other priorities have changed.

During Apollo, the priorities were:

1) Crew Safety,
2) Schedule (land by 1969),
3) Performance (land safely on the Moon),
4) Cost (doesn't matter!)

For an interstellar mission, the public will demand that the priorities be:

1) Crew Safety,
2) Cost (don't break the bank),
3) Performance (launch for the stars),
4) Schedule ("go as you pay")

During the mission, at the highest level, several highly dangerous activities must be accomplished. These include the journey itself, mid-course engine restart (if a ballistic trajectory is chosen) for both mid-course and orbital-insertion de-acceleration, descent to the stellar planet, surface activities, ascent back to orbit from the interstellar planet engine re-ignition for trajectory back to Earth, mid-course engine restart, capture to Earth orbit, and finally landing on Earth.

With these activities, the top two technical priorities that would enable mission success are nuclear propulsion and heavy launch vehicles. These have a direct impact on the number one priority: crew safety. They also affect another critical parameter – minimizing the transit time to the stars.

The transit time needs to be minimized to reduce the human exposure to space. For a mission this long (tens to hundreds of years), we can't expect the crew to remain weightless, then descend and work in a new gravitational environment. Besides having no one to take care of them except themselves, *anything* could go wrong, so the astronauts cannot allow zero-G to affect their performance. Experience with astronauts and cosmonauts has shown that even three hours of exercise a day does not adequately compensate for the debilitating effects of zero-G, so some sort of artificial gravity must be induced, perhaps through rotating part or all of the craft.

Besides the danger of long-term weightless effects, there are dangers associated with cosmic radiation. Two centimeters of aluminum shielding will stop some of the primary solar cosmic rays; however, cosmic rays will produce a "shower" of ionizing particles during the interaction with aluminum. The shower can be absorbed by several centimeters of paraffin, and if the crew's water supply is placed adjacent to the paraffin shield, 16 centimeters of water will absorb the remainder of the shower as well as any other solar cosmic radiation.

As bad as the solar cosmic rays are, galactic cosmic rays are even worse. They would not be stopped by anything except a prohibitively high amount of shielding; however, NASA estimates that the maximum radiation dose due to galactic cosmic rays, although higher than desirable, would still be less than the recommended maximum dose limit of 50 rem.

A major third danger associated with long transit times is the least studied of all, and may turn out to be the most dangerous – the psychological damage that might be incurred when humans are in deep, interstellar space, away from their planet of birth for the first time in history. Because of these and other compelling reasons, the need to shorten the transit time is paramount.

*****

Enabling technologies

Humans will not go to the stars unless they have, as a minimum, nuclear propulsion and heavy launch vehicles. Nuclear propulsion is needed to shorten the transit time to the stars; heavy launch vehicles are needed to boost the enormous amount of material needed for the flight to low-Earth-orbit where the star ship will be assembled.

Chemical propulsion engines are limited to a maximum of 475 seconds of specific impulse. When included in a multidimensional analysis of mass to low-Earth-orbit and transit times, nuclear propulsion is clearly superior to chemical propulsion by at least a factor of two – that is, one hundred percent. Current nuclear propulsion technology, evolving from work performed at Los Alamos National Laboratory in the 1960's on the NERVA program, can yield specific impulses of 925 seconds.

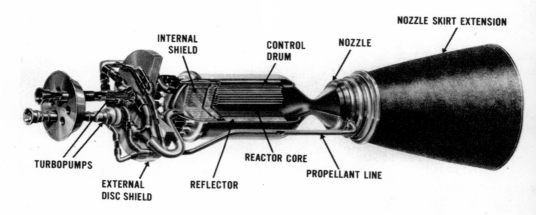

An explanatory drawing of the NERVA (Nuclear Engine for Rocket Vehicle Application) thermodynamic nuclear rocket engine.
Courtesy of NASA

Nuclear electric engines, although having a much higher specific impulse, do not produce as much thrust as thermal nuclear propulsion. Instead of using hydrogen as a propellant, neutral plasma is ejected from the spacecraft. This creates a very low acceleration that can work over many months. The solution may be to use a hybrid system, employing the benefits of both thermal and electric nuclear propulsion. Perhaps the initial "kick" can be attained thermally to boost the star ship into a high velocity trajectory, while the electric propulsion could constantly accelerate the craft onward, to the stars.

The other enabling technology, a heavy lift vehicle, is needed to get the most mass possible to low-Earth-orbit.

Contrary to popular misconception, putting things together in space is not as easy as it seems – especially in a space suit, and especially if you have to suit up (an 8 hour process) and do this many times. Ask any astronaut. They'll tell you that spending up to 4 hours pre-breathing (and another 4 hours post-breathing), getting the skin ripped off their knuckles from tight-fitting gloves, and the sheer strain of extra-vehicular activity is ot the way to do business if you have a choice. For a mission as large and important as going to the stars, it's not wise to start off by making a mega-construction project harder than it needs to be, so the more stuff that can be rocketed to low-Earth-orbit the better.

That's why a main axiom for NASA is to make space flight as simple as possible. Going to the stars demands no less.

Getting the most mass to orbit minimizes putting structures together in space. This not only drives down the cost, but it keeps interfaces to a minimum. Thus, almost paradoxically, although the star ship's nuclear engine may be of a sophisticated hybrid design, the heavy, brute force vehicle needed to initially get material off the surface of the Earth and put it into orbit should be powered by the most robust and simplest engine possible.

One option is to use an updated engine based on Saturn V technology.

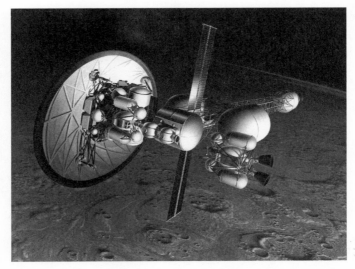

Artist's concept of a nuclear thermal rocket with an aerobrake disk; first investigated as part of the NERVA program, nuclear thermal has more recently been considered for interplanetary human exploration. Courtesy of NASA

A Saturn V launch vehicle mated to an Apollo spacecraft on its way to the launch pad. Courtesy of NASA

Although the old liquid oxygen-kerosene F-1 engine had a relatively low 281 seconds of specific impulse, it still is the largest rocket engine in the world. All 13 Saturn V flights, using a total of 65 engines, were successful.

But whatever engine is selected, it must help power the massive first stage of a heavy lift vehicle capable of boosting at least 250 metric tons to low-Earth-orbit. This mass would greatly simplify initial space operations, decrease the cost of the infrastructure needed to launch the star ship, and provide a more robust means of getting the mission off the ground.

*****

Conclusion

The nation is truly at a threshold – and not a crossroad. A crossroad implies that there exists more than one choice that the nation can make if it is going to get through these tumultuous economic and political times, and flourish. A threshold correctly suggests that *there is no other choice*.

A causal relation exists between long-term investment, a nation's innate need to explore the unknown, and the greatness and survival of a civilization. In 1433, China arguably was the greatest civilization on Earth. Its fleet of merchant and warships docked at Africa, advancing Chinese culture like no other nation before it. But the Ming Empire called back its ships, turned its focus inward on internal problems, withdrew and declined in influence. The Portuguese and the British followed suit … and now standing at a threshold, will America do the same?

The reasons for mounting an interstellar program are many: to gain knowledge of the universe; to advance science; to produce greater technologies for Earth; to focus and force the commercialization of space; to strengthen the US and world economy. But the most important reason is for the survival of the human spirit. This is simply something that humans can't afford *not* to do.

As stated elsewhere, "We must make the decision to either lead, follow, or get out of the way." The bottom line is that if we as humans are going to survive, we *must* follow this path to the stars. There is no other choice.

† Portions of this essay are taken directly from the author's article published in *Analog*, Vol CXII No. 15, mid-December 1992, pp. 68-77.

— PART II —

Anthropology, Genetics & Linguistics

Kin-Based Crews for Interstellar Multi-Generational Space Travel

John H. Moore
Anthropology Department
University of Florida

Over the past several decades, space scientists and writers of science fiction have speculated at length about the optimum size and composition of a crew that could successfully guide a spacecraft from Earth across deep space to another solar system.[1] In these scenarios, space crews in various conditions of animation, representing various combinations of ages, sexes and genders, and often accompanied by assemblages of frozen embryos, egg cells, sperm cells and body parts, and sometimes robots and cyborgs, somehow survive and reproduce over several generations and hundreds of years to reach a habitable destination. Along the way, frozen colleagues are reanimated, egg cells fertilized and embryos implanted in living persons or in laboratory containers to maintain the biological viability of the on-board population. Space crews are not supposed to enjoy the trip, necessarily, but to be brave and maintain their sense of mission in the face of all odds.

Some of the scenarios proposed so far, however, are downright alarming from a social science perspective, since they require bizarre social structures and an intensity of social relationships that are quite beyond the experience of any known human communities. In a recent article in *Discover* magazine, for example, psychologists considered the consequences of the proposed Mars Mission, in which seven heterosexual human adults will purportedly sit shoulder to shoulder for nine months. The title of the article is *Can we get to Mars without going crazy?*.[2] Ominously, Russian Cosmonaut Valery Ryumin is quoted as warning that such a situation fulfills "all the conditions necessary for murder."

Kinship and Demography

In this paper, then, I want to speak as an anthropologist and demographer, and suggest that we are much less likely to go crazy in space, and much more likely to accomplish our interstellar missions, if we send crews into space which are organized according to familiar, ubiquitous, well-ordered, and well-understood social forms. Of the social forms we might try to emulate, some are clearly inappropriate for multi-generational space travel. Roman legions and nunneries, for example, were very efficient for their purposes, but they were not organized to reproduce the species. One particular institution, however, and in fact our oldest and most ubiquitous institution, was organized with reproduction as its central purpose – and that is the human family.

In addition to reproductive efficiency, there are many other good and practical reasons for organizing a space crew along family lines, but let me mention only a few. First of all, human families are structured around biological facts that were, until recently, incontrovertible. Biologically, every human being has a mother and father, and usually a cohort of siblings. Seniority among these kin categories then suggests lines of authority and responsibility. Building on these facts, human societies over the centuries have spun off complex webs of kinship to organize all kinds of social, economic and political activities, from small-scale foraging societies to empires comprising millions of people.

Kinship systems, then, with their inherent seniority between parent and child and between older and younger siblings, can be used to construct a division of labor to accomplish the work of space travel. Robert Heinlein has anticipated this kind of space crew in his description of the Free Traders in *Citizen of the Galaxy*.[3] In Heinlein's book, the eighty Free Traders of the spaceship *Sisu* are organized by sex and seniority, family, clan, phratry and moiety, each of which has specific responsibilities for operating their spaceship and defending it from enemies and competitors. Individuals move up in the hierarchy of work and skill as they get older and more experienced. Figure 1 is a table of organization for the spaceship. The glossary of Table 1 can help understand the vocabulary used by Heinlein, taken from the field of social anthropology.[4]

To sustain and reproduce their spaceship communities, Heinlein's Free Traders maintain a highly structured system of marriage which requires that they occasionally rendezvous with other Free Trader spaceships to court spouses for their young people, and for marriage and other ceremonies.

Overall, our goal in designing an interstellar space community is the same as the Free Traders – to define the parameters and social practices of a bounded but mobile community which is viable, or sustainable in the long term. More specifically and more technically, we want to define a *marital and reproductive regime* for a community which will be capable of traveling, if not indefinitely, at least for a long time alone in space. For present purposes, simply to establish a baseline, we define viability as the ability to travel 200 years, or approximately six to eight normal human generations. Also, we would like to have a space crew or community which is as small as possible, in the interests of economy in building a spacecraft. And, for reasons discussed elsewhere in this volume by Dennis O'Rourke, we would like the space crew to exhibit as much genetic variability as possible.

SPACESHIP SISU

Kinship and Social Organization

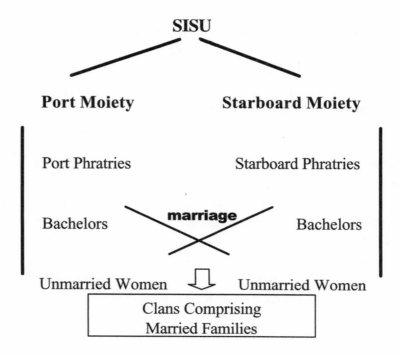

Figure 1 – Table of Organization of spaceship *Sisu*

Table 1 – Glossary of spaceship *Sisu* vocabulary

Age-Set	A group of individuals of the same sex and about the same age
Chromosome	A string of genes. A human has 23 pairs of chromosomes
Clan	A group of families or individuals who are related by descent from a common ancestor
Foraging Society	Comprising small-scale bands who hunt and gather for subsistence
Gene Pool	Alleles (alternative forms of genes) on the chromosomes of a breeding population which are available for reproduction of the next generation
Marital and Reproductive Regime	Rules and traditions governing sexual relations and reproduction, such as incest prohibitions, age of consent, multiple spouses, etc.
Marriage Pool	Number of potential spouses available to a marriageable person in a population
Moiety	A group of phratries or clans constituting half the society
Phratry	A group of clans
Polygyny	A system which allows a man to have multiple wives
Sibship	A cohort of siblings; a group of people sharing the same mother or father, or both

Some Ethical Considerations

Before we begin setting families aboard spacecraft, however, and sending them off toward Alpha Centauri, there are several moral and ethical issues we should consider. Upon first hearing the idea of sending children into space, many people ask, "You would do that to a child?" Their point, of course, is that children sent into space with their parents, have had no opportunity to object. Perhaps a child would rather have spent life as a rock musician or an undersea explorer. Even further, what about children born on board the spacecraft? They will live and die in space with only pictures and videos of Mother Earth.

In answer to such questions, the response heard among astrophysicists is that all of us with children continually make decisions which affect the courses of our children's lives. We change jobs, we move to Chicago, we emigrate to a foreign country. All such decisions constrain the life choices of our children. It is argued that the decision made by parents to join a space crew is not different in kind from decisions made by parents on Earth, it is only different in degree.

Another ethical and practical problem is whether Earth-bound planners, such as ourselves, should be designing social, moral and legal structures to be used by people who will live in an environment so different from ours, and so far away. In the first place, once the space crew is on route, Earth-bound planners can have little or no influence on what happens aboard the spacecraft. If the space crew inaugurates a system of lifetime slavery for some and privilege for others, there is little that the planners on Earth can do to prevent it. Also, we cannot anticipate all the problems that the crew might meet, or design in advance the means to cope with them. In a certain sense, once the spacecraft is on its way, the crew is on its own.

What Kind of Families?

In designing a space crew comprising families, it is necessary to choose some particular kind of family from among the many kinds represented in the whole spectrum of family types present on Earth, simply to maintain homogeneity in spacecraft culture. It would not do to have some families organized as Chinese-style patrilineal clans, while others are organized as polygynous extended families, with each child having multiple mothers. This would confound the arrangements for marriage in the next generation, since different families would have different customs.

In selecting an appropriate family type, also, we must consider some demographic variables which underlie the existence of particular family types on Earth. For example, a system of polygyny requires that there be more women than men in the marriage system, which usually means that some men will be unmarried, since the normal human sex ratio at birth is about 50-50. Each generation, the genes of the unmarried men, some of them unique and not present in other individuals, will be lost to the gene pool of the community.

Monogamy requires about the same number of men as women in the pool of unmarried persons, and has the advantage of maintaining a higher genetic variability than a polygynous system, where a smaller number of men constitute the fathers of the next generation. Small sibships in a marriage system also are beneficial from a genetic standpoint, because they increase the proportion of unrelated persons in the marriage pool. In the long run, this helps sustain genetic variability, and prevents parents with a large number of offspring from dominating the gene pool of the next generation. Postponing parentage as long as possible is also advantageous genetically, because unused chromosomes – even of people with a large number of children – are sloughed off each generation. Therefore a population which experiences only four generations within 200 years will have more genetic variability than a population of the same size experiencing nine generations.

Designing a Marital Regime

All of this demographic engineering carries the danger of undercutting universal human values concerning family life, so perhaps at this point we should make some guarantees to people who join the space crew and to their descendants, a kind of social compact, if you will. To fulfill this compact, we promise to design a marital regime within the following guidelines:

1. Every person will have the opportunity to be married.
2. Each person will be able to choose among at least ten possible spouses with no more than three years of age difference.
3. All potential spouses will be no more closely related than second cousin.
4. Every person will have the opportunity to be a parent.

Limitations of space do not permit describing here all the arguments and considerations which should be taken into account in designing a marital regime for space travel, but let me at least give a general outline.

Figure 2 shows an age-sex pyramid of an expanding human community, a common profile on Earth, but not the kind of distribution we want to see on the spacecraft. Here there are too many children in proportion to adults, and many of the sibships are very large. Especially during the first few years in space, perhaps it would be better to have a higher proportion of adults who could do their work without the distraction of caring for children.

As a final, stable, and sustainable population in space, we want something like Figure 3, a population which is neither increasing nor decreasing, but which fits nicely within the original spatial and geometric contours designed by the spacecraft architects and engineers.

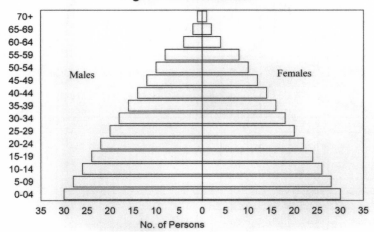

Figure 2 – Age-sex pyramid of an expanding human community

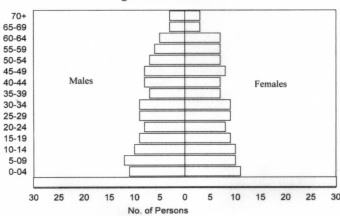

Figure 3 – Static population profile

We can test the performance of this population through time by using a population program called ETHNOPOP©, just developed by my research group.[5] To run a population simulation, you need some vital rates, and we used the following, drawn from the experiences of small-scale societies on Earth. Table 2 is age- and sex-specific, where MD is male deaths, FD is female deaths, and BR is birth rate.

Table 2 – Vital rates used for population performance simulation

Age	MD /100	FD /100	BR /1000	Age	MD /100	FD /100	BR /1000	Age	MD /100	FD /100	BR /1000	Age	MD /100	FD /100	BR /1000
0	20	18		19	2	5	150	38	5	4	50	57	15	11	
1	17	15		20	2	5	150	39	6	4	50	58	15	11	
2	15	12		21	2	5	150	40	6	4	40	59	18	12	
3	13	11		22	2	5	100	41	6	4	40	60	18	12	
4	11	9		23	2	6	100	42	7	5	40	61	18	12	
5	9	8		24	2	6	100	43	7	5	30	62	18	13	
6	8	7		25	2	6	100	44	8	5	30	63	20	16	
7	7	6		26	2	6	100	45	8	5	20	64	20	16	
8	6	5		27	2	6	100	46	9	6		65	20	16	
9	5	4		28	2	6	100	47	9	7		66	20	16	
10	4	3		29	3	6	100	48	10	7		67	30	18	
11	4	3		30	3	6	75	49	10	7		68	40	25	
12	3	2		31	3	5	75	50	12	8		69	50	30	
13	3	2		32	4	5	75	51	12	8		70	60	40	
14	3	2		33	4	5	75	52	15	8		71	70	40	
15	3	2	50	34	4	5	75	53	15	10		72	80	40	
16	2	4	90	35	4	5	75	54	15	10		73	90	40	
17	2	5	150	36	5	5	60	55	15	10		74	100	40	
18	2	5	150	37	5	5	60	56	15	10		⇓	100	⇓	

Testing populations of various initial sizes, we find that a starting population of about 150-180 persons would sustain itself at these rates indefinitely, and stay within the limits of the social compact. But we can do better than this, if we resort to some social engineering. The stable population is not the optimal population, either for an initial crew, or for a crew of constant size, which will continue through space. Although many modifications can be suggested, I will consider here only two of the most promising.

First of all, as a starting population for the space crew, we can use a group of young, childless married couples, instead of a population comprising all ages and both sexes. Ethnographically, this would reflect the actual practice of Polynesian seafaring colonists, who set out in canoe flotillas of young couples to find an unoccupied Pacific island where they might live.[6] The advantage of having a population consisting exclusively of young adults, is that it allows the initial crew to get accustomed and adapted to their new environment before they accept the additional responsibilities of caring for children.

A second modification we can make, in the interests of maintaining genetic variation, is to ask that members of the space crew postpone parenthood until late in a woman's reproductive period. This has two highly desirable consequences, genetically. First, it lengthens the generations, so that fewer chromosomes are sloughed off per century in the population. Second, it results in smaller sibships, since a woman who postponed childbirth until age 35-40 would be unlikely to have more than two or three children. Of course I realize that there are medical consequences for late childbirth, which I cannot discuss here for lack of space.

So what is the result of running a starting group of young couples for several generations, with postponed parenthood? The general contour is shown in Figure 4. Because childbirth is confined to a small number of years, and because the mothers are of nearly the same age, we find that subsequent populations over the next several generations consist of discrete demographic echelons. Using ETHNOPOP to simulate population history for a longer period, we discover that this echelon configuration is very stable through time, because of sexual selection. That is, since humans tend to choose mates of a similar age, the echelons tend to remain bounded and distinct through time, and do not merge or fill the age gaps, even when run for more than two thousand years.

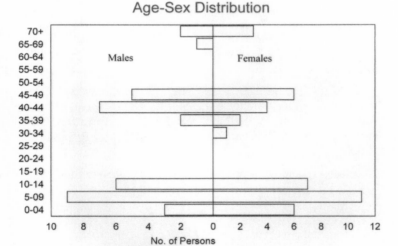

Figure 4 – General contour for postponed parenthood

So, how efficient is this echelon configuration in regard to the promises about marriage and the marriage pool made to potential crew members in the social compact? Running echelon populations of different sizes for hundreds of years, we are surprised to learn that the same promises fulfilled by a normally-distributed crew of 150-180 persons, can be fulfilled by an echelon configuration of about half that size. So it turns out that Robert Heinlein's *Sisu* crew of eighty, if it had been organized into echelons, needn't have made a rendezvous to arrange marriages. They had a community of viable size on board, had it been organized into echelons.

A kind of echelon called an "age-set" is very common among the small-scale societies studied by anthropologists.[7] Potentially, age-sets provide important cultural opportunities for members of the space crew. On Earth, age-sets serve as social and ceremonial organizations. They adopt official mascots, name themselves, devise songs and dances, and define their own groups vis-a-vis other groups. Age-sets are a positive contribution to personal satisfaction and group morale.

Conclusion

In closing, I wish to argue briefly that space travel, especially interstellar, multi-generational space travel, needn't be regarded as an ordeal to be endured by a space crew. Instead, space travel can be organized so that it is pleasant and interesting, as well as full of adventure. It is possible that travelers born in space can expect to experience many of the same pleasures as their relatives on Earth. Under the marital regime outlined above, members of the space crew can expect to have the pleasures of marriage and parenthood, if they want them, and might look forward to having their descendants occupy a space colony or someday return to Earth. In addition, like other prominent people, they will have a place in human history, preserved in the records of their spacecraft. And perhaps best of all, they will have the privilege of designing a new culture, with its own values and beliefs, its own behavior and ceremonies, and its own plans for the future.

1 Mallove, Eugene and Gregory Matloff, *The Starflight Handbook*, 1989, New York: John Wiley and Sons. Berry, Adrian, *The Giant Leap*, 2001, New York: Tom Doherty Associates. Landis, Geoffrey A., "An All-Woman Crew to Mars: A Radical Proposal," *Space Policy* 16 (3): 167-179, 2000.

2 Weed, William Speed (pseudonym), Can we go to Mars without going crazy," in *Discover*, May, 2001, pp. 36-43

3 Heinlein, Robert A., *Citizen of the Galaxy*, 1957, New York: Ballantine.

4 For the basics of kinship, see Fox, Robin, *Kinship and Marriage*, 1967. Harmondsworth, England: Penguin Books. For a basic vocabulary of demography, see Yaukey, David, *Demography: The Study of Human Population*. Prospect Heights, IL: Waveland Press.

5 Moore, John H., "Evaluating Five Models of Human Colonization," 2001, *American Anthropologist* 103 (2): 395-408.

6 Finney, Ben R., "Voyagers into Ocean Space," in *Interstellar Migration and the Human Experience,* Finney and Eric M. Jones, eds, 1985, pp. 164-179. Berkeley: University of California Press.

7 Wilson, Monica, "Nyakyusa Age-Villages," in *Comparative Political Systems*, Ronald Cohen and John Middleton, eds., 1967, pp. 217-227. Garden City, NY: The Natural History Press.

Genetic Considerations in Multi-Generational Space Travel

Dennis H. O'Rourke
Laboratory of Biological Anthropology
University of Utah
Salt Lake City UT 84112

"No amount of intelligent introspection or sophisticated model building can ever substitute for detailed observation and experiment on real life populations."

— *Berry and Peters (1976)*

Each individual is endowed at birth with a genetic constitution. The genetic complement of the cell nucleus is comprised of equal contributions from each parent, while the much smaller, extra-nuclear mitochondrial component is exclusively maternally inherited. The totality of this genetic information is referred to as the individual's **genome** (bolded genetic terms are defined in a glossary, found in Table 1 at the end of the chapter). It is comprised of long strands of deoxyribonucleic acid, **DNA**.

The structure of DNA permits it to transmit hereditary information between generations. Each strand of DNA is actually two complementary strands bound together by hydrogen bonds into the famed double helix. Taken separately, each individual strand of a DNA molecule is composed of a sequence of small, nitrogen based molecules called bases. There are only four such bases: adenine, guanine, cytocine, and thymine. It is the particular arrangement of these bases over long sequences that encode the hereditary information in genes. The four bases are also complementary such that adenine always forms a hydrogen bond with thymine, and guanine always pairs with cytocine. Thus, the sequence of bases that characterizes one strand of a DNA molecule is reflected in its complementary sequence on the alternative strand. For example, the DNA sequence AATGGCTA on one DNA strand would be complementary to the sequence TTACCGAT on the other stand, and each of the complementary bases would be bound together by hydrogen bonds. Any alteration in the original DNA sequence is termed a mutation, and there are many classes of mutations. The substitution of one base for another in a DNA sequence is sometimes called a point mutation, just as breaks in the DNA strand may result in larger rearrangements such as deletions or duplications of DNA sequences. Such changes may have significant effects on the way in which genes encoded in the DNA sequence function, but even if they do not, such mutations contribute to new variation in the genome.

Segments of the DNA that contribute to the production of proteins, say, or other **phenotypic** effects, are called **genes**, and alternative forms of genes are referred to as

alleles. The frequency of alleles in populations helps define the extent and distribution of genetic variation. The evolution of patterns of genetic variation within and between populations of a species is achieved though the action of four primary forces: Mutation, Natural Selection, Gene Flow, and Genetic Drift (Table 2). Anthropological geneticists have long been interested in identifying, and quantifying, the roles of these evolutionary mechanisms in the history of our species. During most of our ancestry, human populations were small, isolated to greater or lesser degrees, and typically highly mobile. They also exhibited relatively slow demographic growth. The rapid growth and large size of many contemporary populations is of very recent vintage. In this context, a small group of explorers, encapsulated in a vehicle for space travel, expecting to be reproductively isolated from other human populations for several generations, is not particularly atypical in our species evolutionary history. Much of what has been learned in the investigation of past, small, 'colonizing' populations may have relevance for the planning of multigenerational space travel.

Table 2 – Evolutionary Mechanisms

Evolutionary Mechanism	Effect
Mutation	Origin of new genetic variants in the genome
Natural Selection	Adaptive force in evolution
Gene Flow	Movement of genes between populations; Promotes intergroup similarity
Genetic Drift	Stochastic[14] change in allele frequencies between generations due to small population size; Promotes intergroup differentiation

It is first useful to briefly deal with each of the evolutionary mechanisms separately, since I believe it permits restricting attention to only two of them. In what follows I assume that the number of space travelers is 150-200 individuals, and that the duration of their travel, and hence their isolation from other human populations, is maximally 6-8 generations.

Mutation is the uncorrected alteration of the sequence of bases in DNA. It provides the initial genetic variation upon which natural selection may act. Mutations are typically deleterious, but, at any specific location in the genome, they occur rarely. Spontaneous point mutations, those involving the misincorporation, or substitution, of one base for another during DNA replication [the copying of DNA during cell division], occur at frequencies of 10^{-9} to 10^{-11} (Cooper, *et al.* 1995). Given an average gene size of ~1.7 kb (kb = kilobase = 1,000 bases), mutations occur in the gene coding region of DNA at frequencies between 1.7×10^{-6} to 1.7×10^{-8} per gene per cell division (Strachan and Read 1999). Thus, given the large number of cell divisions (called mitoses) in the human life span, point mutations at single, specific, sites are rare, but collectively [i.e., across the whole genome] they are common. Most are deleterious, causing death of the cell, although some result in continued and uncontrolled cell division, and are the origin of several cancers. Most mutations of this sort occur in somatic cells and have little import for evolutionary or population considerations. Only those mutations occurring in the germ line (reproductive cells; sperm and ova) can be transmitted to subsequent generations and are, therefore, of interest to population level analyses.

In small populations, over short time frames, mutation is typically unlikely to be an important influence on the genetic composition of a population. However, since the Nobel Prize winning work of H.J. Muller in the 1920's it has been known that radiation causes an elevation in mutation rate (Muller 1927). Ionizing radiation, such as X-rays, gamma-rays, and cosmic rays possess sufficient energy to eject electrons from atoms, creating electrically charged ions. When this takes place in or near molecules of nucleic acids, changes in the base composition of DNA may result.   The sources of ionizing radiation noted above are characterized by wavelengths of up to 150 Å (angstroms) for X-rays to as small as 8×10^{-5} Å for some cosmic rays (Rothwell 1993). However, even the non-ionizing radiation of ultraviolet light (3,800-150 Å) can cause induced mutations.   Experiments on microorganisms indicate that for UV light, wavelengths of approximately 2,600 Å are the most mutagenic (Rothwell 1993). In general, the higher the radiation dose, the greater the effect on mutation rate, whether from UV light or ionizing radiation. In the fruit fly *Drosophila*, a favorite experimental organism of geneticists, the rate of point mutations and single **chromosome** breaks is proportional to the radiation dose.   Double chromosome breaks, which can result in translocation of chromosomal segments from one chromosome to another or inverted sequences within a chromosome, and hence larger, structural alterations to chromosomes, is proportional to the square of the dose, with many more double breaks occurring at high radiation exposures. In general, acute radiation exposure is more effective in increasing mutation rate than chronic radiation exposure.

There are several difficulties in estimating the effects of radiation induced mutations on a population of space travelers. First, mutation is a random event, affecting all regions of the genome equally. Predicting specific outcomes from elevations in radiation induced mutations is not possible in individual cases. Second, many mutations will be somatic (i.e., non-reproductive cells) in origin, resulting in cell death, but in some cases resulting in clinical outcomes of medical concern (e.g., skin cancer). A much smaller proportion will affect germ lines and, therefore, alterations in the genetic composition of subsequent generations. Third, due to the rarity of site specific mutations, human spontaneous mutation rates are estimated with difficulty. As a consequence, estimating the increase of induced mutations as a result of radiation exposure is even more problematic.

An unfortunate natural experiment in this regard is the result of exposure to radiation by surviving citizens of Hiroshima and Nagasaki, Japan following the atomic bombing of those cities in World War II. Over 30,000 children of parents who were within 2,000 meters of the hypocenter of the bombing, and over 40,000 children whose parents were further than 2,000 meters from the hypocenter have been studied for a variety of characters, including untoward pregnancy outcomes, sex of child, malignant tumors prior to age 20, growth and development parameters, **cytogenetic** (chromosomal) abnormalities, and mutations that affect the structure or electrical charge of proteins (Neel 1995). The result was no statistically significant difference between offspring of parents who were maximally exposed to atomic radiation and the offspring of parents who had minimal exposure. Similarly, Kodaira, *et al.* (1995) report no differences in mutation rates in six different DNA segments between children of 50 radiation exposed families in Japan and the children of 50 control families. In contrast, there have been reports of substantial radiation effects on mutation rates as a result of radiation exposure from the Chernobyl nuclear power plant accident (Dubrova, *et al.* 1996).

A full review of mutation mechanisms and radiation effects on them is beyond the scope of this paper. However, it seems clear that space travelers will be exposed to higher levels of both ionizing radiation and UV light than is typical for earth-bound individuals, and therefore, we may expect some increase in the induced mutation rate. Since even small increases in exposure to ionizing radiation, in particular, is associated with an increase in mutation, anticipating the need to shield any craft containing long-term space travelers from the elevated radiation levels found in deep space seems essential.

Natural selection is the dynamic, adaptive force in evolution that operates upon existing population variation. Darwin's (1859) insight into the natural world was that some individuals are more successful at reproduction than others in all populations. Since in sexually reproducing species growth of populations will always outpace increases in necessary resources, individuals within populations are of necessity in competition for the limited resource base. Those individuals whose characteristics (collectively referred to as their **phenotype**) afford some advantage in this competition will benefit, and on average be able to produce more offspring than individuals whose phenotype is less advantageous with regard to resource acquisition. Thus, in each succeeding generation, the frequency of the advantageous phenotypes is expected to increase, and less advantageous phenotypes will decrease. This change in frequency over time as a result of competition in particular environments Darwin termed natural selection. It takes little imagination to realize that to the degree that advantageous phenotypes are underlain by genes that selection of this form can alter the frequency of genes in a directional fashion over time. It is important to recognize that such selection pressures may be highly environmentally specific. Phenotypes that are advantageous in one environment, may be deleterious in another, altering the selective pressures and, accordingly, the distribution of phenotypes and genes over time.

It should be obvious that phenotypic variation is necessary for selection to operate. Without variation adaptation to changing environments is impossible. Since the environment of space travelers is 'artificial', constructed, and controlled, we may assume that it will be designed to approximate the environment on earth from which the travelers come. That being the case, I don't envision any specific alterations in selective pressures from those to which the original crew were already exposed (with the exception of possible increased exposure to ionizing radiation noted above). This may be a faulty assumption, but lacking more specific knowledge of the ship's environment, it seems reasonable to presume that selection will be minimal to absent and ignore its modest effects on the traveling population over the relatively short time-frame of interest (i.e., 6-8 generations).

Gene flow is the transmission of genes from one population to another as individuals migrate. This is not expected to be possible for travelers outside earth orbit, so this mechanism, too, seems irrelevant for our purposes.

Genetic drift is not. Drift is a **stochastic** process whereby genetic material (genes, DNA sequences) are not transmitted to successive generations in the proportions in which they exist in the previous generation. This means that the frequency of alternative alleles at a **locus** (see gene in glossary) may fluctuate at random from generation to generation. This process is only effective in small populations, having little impact on changes in genetic variation in large communities. Since space travelers will constitute a very small population, it is a mechanism for genetic change that merits consideration. The two factors that dictate the magnitude of effect of genetic drift are population size and the number of generations the population remains small. The smaller the population and the longer the time-frame, the more dramatic drift's effect. The effect of drift is simple; it decreases genetic variation in the population. To use a simple example, given enough time and a continuously small population, a locus with only two alleles (bi-allelic) that in the founding generation had equal frequencies of the two alternative alleles, will ultimately become monomorphic. That is, one allele will ultimately become fixed (present at a frequency of 100%), the other will be lost. Which allele becomes fixed and which is lost is entirely random. In the absence of gene flow, the only way the lost allele can be restored is by a new mutation; a very unlikely event in a small, proscribed population.

The effect of genetic drift is to reduce variability in the population by reducing one allele frequency at the expense of the other (ultimately to loss and **fixation**), and in so doing increasing the likelihood of individuals in a population being **homozygotes** at any specific locus. **Heterozygosity** [a measure of genetic variation] is lost. As noted above, the rate at which heterozygosity is lost is dependent on population size and time. The relationship is given, in general, by the following:

$$Het_t \sim 1 - (1 - 1/2N_e + 1)^t \; Het_0$$

where *Het* is heterozygosity, N_e is effective population size, and t is time in generations (Crow 1986).

As illustrated in Table 3, if we let effective population size (N_e) be either 150 or 200, the number of generations (t) be either 6 or 8, and the initial heterozygosity in the founding generation (*Het_0*) be 0.5, 0.75 or 1.0, the reduction in heterozygosity ranges between approximately 1% and 3%. The action of drift is similar to that of inbreeding, discussed by Moore (this volume).

Table 3 – Effects of genetic drift on heterozygosity with varying effective population size (N_e), generational time (6 or 8 generations), and initial heterozygosity estimates (*Het_0*)

N = 150	N_e = 200
$Het_0 = 0.5$ - $Het_6 \approx .4901$	$Het_0 = 0.5$ - $Het_6 \approx .4926$
$Het_0 = 0.75$ - $Het_6 \approx .7352$	$Het_0 = 0.75$ - $Het_6 \approx .7388$
$Het_0 = 1.0$ - $Het_6 \approx .9802$	$Het_0 = 1.0$ - $Het_6 \approx .9851$
$Het_0 = 0.5$ - $Het_8 \approx .4869$	$Het_0 = 0.5$ - $Het_8 \approx .4901$
$Het_0 = 0.75$ - $Het_8 \approx .7303$	$Het_0 = 0.75$ - $Het_8 \approx .7352$
$Het_0 = 1.0$ - $Het_8 \approx .9737$	$Het_0 = 1.0$ - $Het_8 \approx .9802$

Inbreeding results from the mating of individuals related by ancestry, who by definition share more alleles (alternative forms of a gene) in common at individual genetic loci than expected by chance. The offspring of such couples are therefore homozygous at loci more often that expected under random mating among unrelated individuals. This increase in homozygosity (proportion of loci that are homozygous) among the offspring of related individuals is related to the degree of relationship between the parents, but like genetic drift, increasing the frequency of homozygotes decreases overall heterozygosis, and, consequently, the extent of genetic variation. Inbreeding is measured as the probability of **autozygosity** by the Inbreeding Coefficient:

$$F_t = 1 - (1 - 1/2N_e)^t$$

where, as before, N_e is effective population size, and t is number of generations (Crow 1986, Hartl and Clark 1989). Table 4 indicates that the increase in autozygosity under the conditions assumed above ranges between approximately 1.5% and 2.5%, similar to the value obtained for drift effects .

Table 4 – Changes in probability of autozygosity (F-value) with
varying effective size (N_e) and generational time

$N_e = 150$	$N_e = 200$
$F_6 = 0.0198$	$F_6 = 0.0149$
$F_8 = 0.0263$	$F_8 = 0.0198$

The reduction in variation due to drift or inbreeding described here appears minor. Indeed, a reduction of 2-3% in a population's level of genetic diversity is not particularly alarming. Many successful colonizing or bottlenecked populations in human history have experienced substantially greater reductions in variation. If we envision the space travelers returning to Earth after 6-8 generations, and re-integrating into Earth's population, then the reduction in genetic variation anticipated is transient, a short-term phenomenon. In an evolutionary context, then, the genetic effects are not particularly troublesome, and need not cause undue concern in planning and policy considerations. This is not to suggest that genetic considerations are irrelevant to such planning, however. The genetic relevance is rather more practical, and focuses on the genetic realities of short-term effects in small populations.

The practical realities of drift and inbreeding (and they act simultaneously in small populations), is an increase in individuals who are homozygous at multiple loci. And it is the increase in homozygosis that is of immediate concern. It has been hypothesized that each individual carries up to 5 recessive lethals. That is, each individual carries multiple alleles in heterozygous loci that if paired with another such allele would result in death. Increases in homozygosity due to drift or inbreeding, then, would be expected to result in an increase in mortality, but perhaps more importantly, an increase in disease phenotypes that are the result of recessive alleles. For recessive alleles two copies of the allele are required (i.e., a homozygote) for the recessive phenotypic to be expressed. This effect is more pronounced for rare alleles than for more common ones. For example, approximately 1 in 25 people of European ancestry are heterozygous for the primary cystic fibrosis mutation. Mating with a first-cousin increases the likelihood of an affected offspring by a factor of 3 over mating with an unrelated individual. In contrast, only about 1 in 170 individuals are heterozygous for galactosemia, a disorder of carbohydrate metabolism, and first-cousin matings increase the probability of an affected (homozygous) offspring by a factor of 21 (Jorde, et al. 1999).

Indeed, increased (pre-reproductive) mortality among the progeny of related parents has been observed in several populations around the world (Table 5). The average increase in mortality among offspring of second-cousin marriages is ~2.2% in these data (2.7% excluding the Utah sample).

Table 5 – Mortality rates (in %) based on degree of parental relationship
(After Jorde, *et al.* 1999)

	1st Cousin	2nd Cousin	Unrelated
Amish	14.4	13.3	8.2
India	11.7	5.9	5.8
France	17.7	11.7	8.6
Japan	14.5	12.0	10.4
Pakistan	22.1	20.1	16.4
Sweden	14.1	11.4	8.6
US (Utah)	22.4	12.2	13.2

I don't have comparable figures for debilitating recessive phenotypes that are not fatal (i.e., increase morbidity but not necessarily mortality), but we can expect them to increase in comparable frequencies. Thus, we may anticipate some increase in medical conditions that require additional levels of care, including developmental delays and abnormalities, mental retardation, neurological disorders, etc. Many such phenotypes will have variable expressivity, and may entail considerable investment of time and resources for many years per affected individual. If many such patients are born in a closed, small population, such as that anticipated for multigenerational space travel, resources may be strained.

There are ways of minimizing such cases, of course. Many of the known deleterious recessives can be identified in carriers, and an expanded panel of genetic tests could be employed to exclude carriers as participants in such a multigenerational space exploration effort. In addition to the classical recessive diseases (PKU, cystic fibrosis, galactosemia, Wilson's disease) there is an emerging field of investigation into **triplet repeat** diseases. In these disorders, a triplet of DNA nucleotides is repeated many times in tandem, resulting in a disease phenotype (e.g., Huntington's disease, myoclonus epilepsy, myotonic distrophy, and several ataxias). Should potential travelers in generation 1 be screened for these repeats so as to exclude those at increased risk of transmitting an expanded repeat series to an offspring?

One other practical matter may be relevant. Moore's simulations (this volume) suggest a stable population may be obtained, and an 'age-set' structure established by delaying reproduction. This is undoubtedly true. However, some phenotypes, for example trisomy 21 (Down's syndrome), achondroplasia, and Marfan syndrome (Jorde, *et al.* 1999), increase in frequency with advancing parental age. Delaying reproduction may have advantageous demographic characteristics, but these advantages may be countered by a corresponding increase in the frequency of conditions requiring increased medical care, attention, and resources.

It has been suggested that the optimal crew for multigenerational space travel be exclusively female (Landis 2000), perhaps to be accompanied by an onboard sperm bank. Given the state of the science in in vitro fertilization technology (although this is beyond my own expertise), this may be technically possible. It would even be possible to regulate the sex of offspring produced by such methods. In the same way, screening donors and recipients for genetic characters is also technically possible, and might be employed to minimize the appearance of specific phenotypes in the isolated, space bound population.

I have focused on genetic screening to prevent the occurrence of specific phenotypes, but such testing may also be used as a positive selection mechanism to enhance, or at least increase, the likelihood of certain phenotypes. Depletion of bone mass is a specific problem in long-term exposure to microgravity. Bone mineral density varies among individuals, and average bone mineral density may vary between groups. Given the success of the Human Genome Project in providing a first sequence of the entire human nuclear genome, coupled with recent developments in genomics, bioinformatics, and quantitative genetic analytical methods, it is not unreasonable to predict that by the time a group of individuals is selected to begin a multigenerational space voyage, the genes affecting increased bone mineral density may well be known at the molecular level (e.g., Johnson, et al. 1997, Koller, et al. 1998, Gong, et al. 2001). Should only people who possess 'high bone mineral density alleles' be included in the crew? Should offspring born on the voyage be screened prenatally to assure they, too, possess the 'desired' characters? Natural selection may be of little import to the scenario of space travel over a few generations, but artificial selection may not.

In addition to bone mineral density, or calcium absorption, a related characteristic that is also likely under genetic control, additional characters that may be thought of as 'Enhanced Health Phenotypes' may also be identified for genetic screening. The dopamine receptor DRD4 has received much attention lately, with some investigators suggesting that polymorphism at this locus is associated with 'novelty seeking' (e.g., Benjamin, et al. 1996, Ebstein, et al. 1996, but see Gelernter, et al. 1997). It has been hypothesized to have arisen and realized a selective advantage through the spread of early human populations (e.g., Ding, et al, 2002, Harpending and Cochran 2002). In contrast, the proposed 'novelty seeking' allele of early human populations is now associated with elevated risk to attention deficit with hyperactivity disorder (ADHD) in modern children (Faraone, et al. 2001). Should this genetic marker be screened, and decisions made regarding inclusion in generation 1 of the space travelers, or even used in reproductive decisions in the generations of travel? Which will be viewed as the important aspect of this hypothesized behavioral genetic effect? The novelty seeking thought to be advantageous to migrating and exploring ancestors, or the attention deficit in closed school rooms of today?

How much genetic technology and knowledge is to be employed to structure th
gene pool of subsequent generations in multigenerational space travel seems to me to b
the most immediate and critical problem to resolve. Such a scope for genetic screenin
is unprecedented and would likely raise serious ethical, legal, and social questions tha
would need to be dealt with and resolved during the planning stages of the projec
These ethical and social issues are not new. They are, in general, the same ones bein
debated today with respect to individual testing, population genetic screening, an
medical intervention in reproductive decisions. To the degree they are expanded i
scope and application in our hypothetical space traveling isolate, and to the degree tha
they might enhance the group's genetic divergence from earthbound populations, the
take on a sharper focus.

Table 1 – Definition of genetic terms used in text

Allele	alternative form of a gene
Autozygous	homozygote whose similar alleles at a locus are identical by descent
Cytogenetic	Oo or relating to chromosomes, and chromosomal abnormalities
Chromosome	coiled DNA strands bound to proteins (histones); visible microscopically
DNA	deoxyribonucleic acid; a double-stranded, helical molecule; each strand is composed of a sugar-phosphate backbone bound to a sequence of four nitrogenous bases (A, T, G, C); complementarity of these bases (A to T and G to C) gives DNA its double stranded, helical structure
Fixation	setting an allele frequency to 1.0 (=100%)
Gene	hereditary unit; specific position (locus) in the genome; a specific stretch of DNA sequence influencing expression of a phenotype
Genome	totality of the genetic material in a cell
Genotype	specification of both alleles present at a specific locus in an individual
Heterozygosity	a measure of genetic variation; for a given locus it is the probability that two randomly selected alleles in a population are different; in a randomly mating population it is also the expected proportion of heterozygotes at a locus given the observed allele frequencies
Heterozygote	the genetic constitution of an individual who inherits two dissimilar alleles (one from each parent) at a locus; a genotype comprised of two different alleles
Homozygote	the genetic constitution of an individual who inherits similar alleles (one from each parent) at a locus; a genotype comprised of two similar alleles
Mutation	any change in DNA sequence; a process resulting in a structural change to a chromosome
Phenotype	observable characteristic(s) of an organism; produced by the underlying genotype in conjunction with the environment
Stochastic	a random process or event
Trinucleotide Repeat	the tandem repetition of three DNA bases in the genome; typically, repeat number determines phenotype

References

○ Benjamin J, L Li, C Patterson, *et al.* (1996) Population and familial association between the D4 dopamine receptor gene and measures of novelty seeking. *Nature Genet.* 12:81-84.

○ Berry RJ and J Peters (1976) Genes, survival and adjustment in an isolated population of the house mouse. IN: S. Karlin & E. Nevo, eds. "Population Genetics and Ecology". New York: Academic Press, Inc.

○ Cooper DN, Krawczak M and Antonorakis SE (1995) The nature and mechanisms of human gene mutation. IN: C. Scriver, AL Beaudet, WS Sly and D Valle, eds. "Metabolic and Molecular Bases of Inherited Disease, 7th ed." New York: McGraw-Hill.

○ Crow JF (1986) "Basic Concepts in Population, Quantitative, and Evolutionary Genetics." New York: W.H. Freeman & Co.

○ Darwin C (1859) "The Origin of Species by Means of Natural Selection or The Preservation of Favoured Races in the Struggle for Life." London: Murray.

○ Ding Y, H-C Chi, DL Grady, *et al.* (2002) Evidence of positive selection acting at the human dopamine receptor D4 gene locus. *Proc. Natl. Acad. Sci., USA* 99:309-314.

○ Dubrova YE, VN Nesterov, NG Krouchinsky, *et al.* (1996) Human minisatellite mutation rate after the Chernobyl accident. *Nature* 380:683-686.

○ Ebstein RP, O novick, R Umansky, *et al.* (1996) Dopamine D4 (DRD4) exon III polymorphism associated with the human personality trait of novelty seeking. *Nature Genet.* 12:78-80.

○ Faraone SV, AE Doyle, E Mick and J Biederman (2001) Meta-analysis of the association between the 7-repeat allele of the dopamine D4 receptor gene and attention deficit hyperactivity disorder. *Amer. J. Phychiatry* 158:1052-1057.

○ Fisher SE, C Francks, JT McCracken, *et al.* (2002) A genomewide scan for loci involved in attention-deficit hyperactivity disorder. *Amer. J. Hum. Genet.* 70:1183-1196.

○ Gelernter J, H Kranzler, E Coccaro, *et al.* (1997) D4 dopamine-receptor (DRD4) alleles and novelty seeking in substance-dependent, personality-disorder, and control subjects. *Amer. J. Hum. Genet.* 61:1144-1152.

○ Gong Y, RB Slee , Naomi Fukai, et al (2001) LDL Receptor-Related Protein 5 (LRP5) Affects Bone Accrual and Eye Development. *Cell* 107:513-523.

○ Harpending H and G Cochran (2002) In our genes. *Proc. Natl. Acad. Sci., USA* 99:10-12.

○ Hartl DL and AG Clark (1989) "Principles of Population Genetics." Sunderland, MA: Sinaur Assoc.

○ Johnson M.L., Gong G., Kimberling W., *et al.* (1997) Linkage of a gene causing high bone mass to human chromosome 11 (11q12-13) *Amer. J. Hum. Genet.*, 60:1326-1332.

○ Jorde LB, JC Carey, MJ Bamshad and RL White (1999) "Medical Genetics." St. Louis: Mosby.

○ Kodaira M, C Satoh, K Hiyama and K Toyama (1995) Lack of effects of atomic bomb radiation on genetic instability of tandem-repetitive elements in human germ cells. *Amer. J. Hum. Genet.* 57:1275-1283.

○ Koller D.L., Rodriguez L.A., Christian J.C., *et al.* (1998) Linkage of a QTL contributing to normal variation in bone mineral density to chromosome 11q12-13. *J. Bone Miner. Res.*, 13:1903-1908

○ Landis GA (2000) An all-woman crew to Mars: A radical proposal. *Space Policy* 16:167-179.

○ Muller HJ (1927) Artificial transmutation of the gene. *Science* 66:84.

○ Neel JV (1995) New approaches to evaluating the genetic effects of the atomic bombs. *Amer. J. Hum. Genet.* 57:1263-1266.

○ Strachan T and AP Read (1999) "Human Molecular Genetics", 2nd ed. New York: Wiley-Liss, Inc.

Language Change and Cultural Continuity on Multi-Generational Space Ships

Sarah G. Thomason
University of Michigan

My assigned task in this symposium is to consider problems related to linguistic and cultural continuity that the space travelers will have to deal with. I'll start with language, and with the assumptions (suggested by John Moore) that the new community will comprise about 200 people and the journey will take about 200 years.

The most basic question is, what language(s) should the travelers speak? First, they should all speak the same language, for obvious reasons. If they can't communicate with each other to begin with, it will take a lot of time and effort before they learn how to do so, and the language learning or creating process will distract them from the many other things they'll need to do to construct their new society. This problem could perhaps be avoided by starting with, say, two adults who speak the same language and 198 newborn infants; the infants would then grow up speaking the adults' language and/or creating their own new language, which would be a creole language. They would have their shared language up and running in a very few years, because children are terrific language learners. But this doesn't seem like a practical scenario, so let's start with 200 people who share a single language.

One might suppose that the travelers should speak "a simple language," one that lacks the irregularities and other complications of most communities' languages – maybe an artificial, deliberately designed language, like Esperanto. Unfortunately, linguists have found that there's no such thing as a simple language. Any language that is widely used for ordinary every-day communication is inevitably going to be complicated. Even Esperanto, which was designed to be regular and (in some ways) quite simple, has changed and acquired the same kinds of complexities as natural languages, as its use has spread. But Esperanto wouldn't in any case be the ideal choice for the space vehicle's language. Although it has numerous speakers on several continents (maybe all continents except Antarctica), the pool of speakers is probably too small to permit enough flexibility in choosing the space travelers.

I would suggest that English is the obvious and best choice for the space vehicle's language. This choice isn't based on jingoism, but on practicality. To achieve genetic diversity in the space vehicle's population, the organizers of the journey might wish to attract volunteers from a variety of populations. English is, and has been for some ecades now, the world's major international language, as indicated by (among other hings) its exclusive use by air traffic controllers at international airports and its very xtensive use in higher education and business all over the world. As of 2000, fifty-one ations had English either as their only official language or as one of two or more fficial languages. If we count the United States, where English is certainly dominant ven though the U.S. has no official language, that makes fifty-two countries. Four other ations have English as a semi-official language. (There were 195 nations in the world n 2000.) The countries with official English are scattered widely throughout the world, n Oceania and other island nations as well as on all continents except Antarctica and outh America. In many of those countries, only educated people actually speak Inglish; but it seems likely that volunteers for the space trip will be relatively well ducated, and this therefore isn't likely to be a serious difficulty.

It might even be possible to find enough genetic diversity in the U.S. alone, given he country's melting-pot history, not to mention the genetically diverse Native American peoples who were here before European colonization, and who now speak nainly English. But it would be easier to draw on volunteers from a variety of countries, nd it wouldn't be hard to find genetically diverse English speakers in such countries as Australia (especially among the Aborigines), Singapore, Ghana, and India, in addition o the U.S. (European Americans, Native Americans, and immigrants from different arts of the world).

The question of language change arises. If we assume that the entire trip will take about 00 years as observed in the space vehicle, we probably don't have to worry much about najor changes in language structure that would make the travelers who return to Earth hard o understand. The reason is that language structure tends to change slowly enough that 200 ears is not long enough to effect drastic restructuring: it's still possible, for instance, for nodern English speakers to read Shakespeare, who wrote about 400 years ago.

But we might well have to worry about vocabulary change, because the travelers' nvironment will be so different from any Earthly setting that new words will inevitably pring up and many old words will fall out of use. Basic vocabulary like mother, father, un, walk, sit, water, I, you, one, two, three, animal, and the like will no doubt persist; ut words like snow, windy, river, ocean, mountain, sunburn, summer, winter, horse, iger, and ostrich are likely to be non-useful in the travelers' world, as well as words for on-portable Earth-bound cultural artifacts like car, train, boat, truck, airplane, kyscraper, tunnel, bridge, and so forth. Moreover, teenagers growing up on board the pacecraft will presumably be like teenagers everywhere in their desire to have their own ocabulary to use as an in-group "language." The words they create are unlikely to be he same as words invented by a comparable group of teenagers on Earth. Some of their ew words will surely remain in their language when they grow up (think of words like

"cool" in modern American adult speech, and old slang words like "mob" that have lon‍
since lost their slangy flavor). Still, the lexical divergence between the space traveler‍
and English speakers on Earth should not be as great as the divergence betwee‍
Shakespeare's vocabulary and modern American English vocabulary, so the traveler‍
who return to Earth should be able to talk to English speakers on arrival.

A caveat is necessary, however. It happens occasionally on Earth that a sma‍
speech community will deliberately alter the community's language to make ‍
purposely different from neighbors' languages. The reasons are social, not linguistic ‍
it happens when speakers want to differentiate themselves more sharply from thei‍
neighbors, and language is such a potent symbol of identity that changing or disguisin‍
the language is an effective means of differentiation. It seems quite possible that th‍
space travelers, bonding together as a unique community, will want to distinguis‍
themselves linguistically from the people they left behind. In addition, if volunteers ar‍
drawn from different countries, they won't start out speaking the same dialects o‍
English: their varieties of the language will differ in sound structure, sentence structure‍
and perhaps even word structure.

Most of this variation will be relatively minor; still, resolving it into a singl‍
relatively homogeneous dialect – which is likely to happen with the first generation o‍
children born on the space vehicle – will surely result in a dialect that differs from all o‍
the parents' dialects, and from every other dialect of English spoken on Earth. (Wel‍
unless most of the volunteers are in fact from the U.S., in which case they'll probabl‍
all speak very similar dialects and may be able to impose their own dialect on their non‍
American companions and their children.)

Let's turn now to the topic of cultural continuity, specifically continuity betwee‍
the Earthly culture(s) left behind and the emerging new culture on the space vehicle‍
How much continuity should be aimed for, and how should it be achieved? The worl‍
of the space vehicle will be so different from Earth that, just as many words wil‍
suddenly be useless, many cultural activities and norms (even assuming that those ar‍
shared by all the volunteers!) will be non-useful, inoperative. You can't ski aboard ‍
spacecraft, or go river-rafting, or hike in the mountains, or watch live football games o‍
television, or go to a charity ball, or volunteer in a soup kitchen; you can't commute lon‍
distances to work, or choose from the wide variety of occupations available on Earth, o‍
watch breaking news in real time. You can't travel to new and different places inside th‍
spacecraft, or meet any people other than the on-board population. And those are jus‍
the American-style activities.

What about religion? If the population is drawn from several different countries‍
or even if they all come from the U.S., it'll be very hard to ensure that they have the sam‍

religious beliefs and practices. Some measures must be taken to avoid religious strife onboard. The most obvious, perhaps, is to administer questionnaires and personality tests to the hordes of volunteers who will no doubt present themselves as candidates for the space vehicle's crew, and to attempt to choose open-minded people without a trace of religious fanaticism: no fanatics, no religious strife (with luck). There is no way to prevent religious tensions from arising on board, but since the community must remain small, sects may be unlikely to develop.

Choosing all the travelers from a single religious persuasion might be wise, if it's feasible; but this goal would probably be secondary to the goal of (for instance) ensuring genetic diversity.

The crucial question remains: for all those activities and social norms that aren't going to transfer into the space vehicle, how much continuity is (a) desirable and (b) attainable? Teaching children about life on Earth in classrooms is likely to engender about as much enthusiasm as teaching (say) American schoolchildren about World War I in classrooms. It'd help, maybe, if TV programs and movies are taken along on the trip – tapes of "Bart Simpson" would convey information about some U.S. social norms (though they might not appeal to non-Americans on board). So, the first problem is to decide what cultural features should be focused on in the space vehicle's continuity program.

This is not an easy question to answer, even if all 200 people onboard come from the very same cultural background (which they won't, presumably). It'd be hard to explain sailing or river-rafting or ice-skating to people who will never have a chance to try out such activities, or even to see a large body of frozen or unfrozen water. But many people on Earth don't do any of those things either, so maybe the space travelers could just be allowed to forget them and not teach them to new generations. Other things, though, would probably seem important to the travelers: the geography of Earth, recent history of Earth (including scientific breakthroughs and major wars and natural disasters and maybe even major musical and other artistic events), the nature and history of world religions, major cultural achievements (they can't visit the Louvre or a symphony hall, but they can see images from the former and hear music from the latter, just as many people on Earth do).

Once the features have been selected for transmission, the media for transmission must be chosen: books, lectures, videos, recorded music, etc. There will, one assumes, be musical instruments onboard the spacecraft, so that the travelers can make music themselves, both playing old compositions and composing new music; and likewise for painting materials, drawing materials, and other artistic media. For entertainment, and probably for education as well, there will be computer games.

— PART III —

Looking for Life in Unlikely Places

Looking for Life in Unlikely Places: Reasons Why Planets May Not Be the Best Places to Look for Life[†]

Freeman J. Dyson,
Institute for Advanced Study,
Princeton, New Jersey

Abstract

A new method is proposed to search for extraterrestrial life adapted to cold environments far from the sun. To keep warm, using the light from a distant sun, any life form must grow a system of optical concentrators, lenses or mirrors, to focus sunlight onto its vital parts. Any light not absorbed, or any heat radiation emitted from the vital parts, will be focused by the optical concentrators into a narrow beam pointing back toward the sun. To search for such life forms, we should scan the sky with optical and infrared telescopes pointing directly away from the sun. Any living vegetation will be seen as a bright patch in strong contrast to its dark surroundings, like the eyes of a nocturnal animal caught in the headlights of a car. This method of search may be used either with space-based or with ground-based telescopes. Examples of places where the method would work well are the surfaces of Europa, Trojan asteroids, or Kuiper Belt objects. Any life form that adapted successfully to a vacuum environment would be likely to spread widely over objects with icy surfaces in the outer regions of the solar system.

1. Pit-lamping at Europa

I leave it to the audience to decide whether this chapter is an elaborate hoax, a piece of bad science fiction, or a serious contribution to the planning of future space missions.

My son, who lived for many years in Canada, informs me that in rural Canada, night-time hunting with a carefully shielded flashlight is called "pit-lamping." It is illegal, because it is too effective at turning up game. The reason why it is effective is that the eyes of animals staring into the light act as efficient reflectors. The eye reflects a fraction of the light that is focused onto the retina, and the reflected light is again focused into a narrow beam pointing back to the flashlight. If you are standing behind the flashlight, you can see the eyes as bright red points even when the rest of the animal is invisible. Pit-lamping is too easy and too efficient. It results in extermination of the game, spoiling the fun for other hunters. The same strategy is used legally by sheep-farmers in New Zealand

to exterminate rabbits who compete with sheep for grazing land. The farmers drive ove
the land at night in heavily armed land-rovers with headlights on, shooting at anythin
that stares into the headlights and does not look like a sheep.

I am suggesting that pit-lamping would be a good strategy for us to use when we ar
searching for life in the solar system. In this talk, I will not discuss the search fc
intelligent life. The search for intelligent life is a fascinating subject, but it is not on th
agenda for today. Today I am discussing the search for dumb life, the sort of life tha
existed on this planet before humans came along. Consider for example the problem c
looking for life on Europa. If life exists on Europa, it is most likely that it exists in th
ocean under the ice. The ice is many kilometers thick. To look for life in the ocean wi
be very difficult and expensive. It would be much easier and cheaper to look for life o
the surface. Life on the surface is less likely to be there, but easier and cheaper to detec
When we are choosing a strategy to look for life, detectability is more important tha
probability. Estimates of the probability of life existing in various places are grossl
unreliable, but estimates of detectability are generally quite reliable. Probability depend
on theories of the origin of life about which we know nothing, while detectability depend
on capabilities of our instruments about which we know a lot. So I state as a genera
principle to guide our efforts to search for life: go for what is detectable, whether or nc
we think it is probable. For all we know, all kinds of unlikely things might be out there
and we have a chance to find them if and only if they are detectable.

When we are looking for life on Europa, we must begin by considering th
question: what kind of life could exist and survive on the surface of Europa, where th
ambient temperature is about 120 degrees Kelvin or minus 150 Celsius? We ma
imagine that life originated and evolved for a long time where it is warm, in the dar
waters below the ice. Then by chance some living creatures were carried upward
through cracks in the ice, or evolved long shoots pushing up like kelp through the crack
and so reached the surface where energy from sunlight was available. In order to surviv
on the surface, the creatures would have to evolve little optical mirrors concentratin
sunlight onto their vital parts. Some arctic plants on planet Earth have already evolve
something similar, with parabolic petals reflecting sunlight onto the ovaries and seed
that grow at the center of the flower. On Europa the mirrors would have to be mor
powerful than on the arctic tundra of Earth, but not by a big factor. Sunlight on Europ
is 25 times weaker than sunlight on Earth, so the mirrors on Europa would have t
concentrate sunlight by a factor of 25. That would be sufficient to maintain th
illuminated area at the focus of the mirrors at a temperature of 300 Kelvin, at whic
water is liquid and life as we know it can function. The mirrors would not need to b
optically perfect. Concentration of sunlight by a factor of 25 could be done with a roug
approximation to parabolic shape.

Now suppose that we have a Europa orbiter looking for life on Europa. If the life happens to be on the surface, it must have some kind of reflectors to concentrate sunlight. Then all we have to do is arrange for the orbiter to scan the sunlit face of Europa, always pointing the camera in a direction directly away from the sun. If there is an optical concentrator on the surface, any sunlight that is not absorbed at the focus will be re-emitted and reflected into a beam pointing back at the sun. Like the eyes of a moose in Canada or the eyes of a rabbit in New Zealand, anything alive on the surface will then be 25 times brighter than its surroundings, assuming that the illuminated surface at the focus of the concentrators has the same albedo as the surroundings. Even if the surface at the focus has lower albedo, so long as the albedo is not close to zero, any patch of life on the surface will stand out from its surroundings like the animals in a pit-lamping hunt. Since our purpose is not to kill and eat the creatures but only to discover them, there is no reason why pit-lamping on Europa should be illegal.

If any of you have sat in a window seat on the shady side of a commercial airliner and looked down at a layer of cloud not too far below you, you will probably have seen the beautiful bright halo that sometimes surrounds the shadow of the airplane as the shadow moves along the clouds. The halo is produced by back-scattering of light in water droplets or ice-crystals in the cloud. This is only one of many ways in which non-living materials can mimic the back-scattering effect of a living creature. The fine dust on the surface of our Earth's moon also produces a sharp increase in albedo when the Moon is observed with the sun directly behind the Earth. If we observe strong back-scattering of sunlight from a patch on the surface of Europa, we should not immediately claim that we have discovered life. Other explanations of the back-scattering must be carefully excluded before any claims are made. A good way to do this would be to carry a sensitive infrared bolometer looking in the same direction as the optical camera. If the optical back-scattering is caused by reflection from optical concentrators attached to living creatures, then the warm surface at the focus of the concentrators will radiate thermal infrared radiation, and the thermal radiation will also be reflected back in a narrow beam pointed toward the sun. The infrared bolometer will detect a signal from a patch of living creatures viewed from the direction of the sun, and this signal will stand out from the cold surroundings even more strongly than the optical signal. It is unlikely that any non-living back-scatterer could mimic this infrared signal.

Another way to distinguish living from non-living back-scatterers would be to analyze the back-scattered light with a spectroscope and look for spectral features that might be identified with biologically interesting molecules. Laser illuminators of various wavelengths could also be used. But an infrared bolometer, if it is available and not too expensive, would give a more reliable diagnosis of reflectors as living or non-living than an optical spectroscope.

The same strategy of orbital pit-lamping could be used to search for evidence of life on any celestial body that has a cold surface with a transparent atmosphere or no atmosphere at all, and is situated far from the sun but not too far for us to send an orbiter. After we are done with Europa, we could try the same technique on Ganymede and Callisto, on any of the moons of Saturn except Titan, on the moons of Uranus and Neptune, on the Trojan asteroids, or on asteroids in the main belt between Mars and Jupiter. But this would not be a very promising way to go, if the only purpose of the orbiters were the search for life. After the first two or three orbiters had failed to find evidence of life, it would be hard to get funding for more of the same. It should be a fundamental principle in the planning of space operations, that every mission searching for evidence of life should have other exciting scientific objectives, so that the mission is worth flying whether or not it finds evidence of life. We should never repeat the mistake that was made with the Viking missions, whose advertised purpose was to give a definitive answer to the question whether life exists on Mars. After the Viking missions failed to find evidence of life, the further exploration of Mars was set back for twenty years. We should be careful to see that every orbiter dispatched to the outer solar system takes full advantage of the opportunity to explore a new world, whether or not the new world turns out to be inhabited. Fortunately, there is another promising and cost-effective way to search for life beyond Europa. The second way is to use pit-lamping, not from orbit close to the target but from back here on Earth. The second kind of pit-lamping, which I call home-based pit-lamping, can be done from telescopes like Hubble close to the Earth, or even better from big telescopes on the ground.

2. Home-based Pit-lamping

When we do home-based pit-lamping, we make a big jump in the distance and the number of objects that we might hope to search. Think of the Kuiper Belt, which begins about ten times further from the sun than Europa and is supposed to contain billions of objects of size ranging from a kilometer upward. Suppose that life has somehow succeeded in establishing itself on the surface of one of these objects. This may seem unlikely, but nobody can prove it to be impossible. Let us imagine that the surface of one of the objects at the inner edge of the Kuiper Belt is covered with living creatures. These creatures are living in sunlight that is a hundred times weaker than sunlight on Europa. So they must have evolved optical concentrators that are a hundred times stronger. They must concentrate sunlight by a factor of 2,500 instead of 25. This still does not require high-precision optics. The angular resolution of the concentrators must be of the order of one degree instead of ten degrees. This is still about fifty times less precise than the optics of the human eye. A roughly parabolic reflecting surface would be good enough to do the job. When sunlight falls upon the concentrator, the light that is not absorbed is reflected into a beam one degree wide that is pointed back toward the sun. Now it happens that the Earth's orbit around the sun is also about one degree wide hen seen from the inner edge of the Kuiper Belt. The Earth is close enough to the sun, o that the reflectors will increase the brightness of the object as seen from the Earth by factor of the order of 2,500. As the Earth moves around the sun, when we look outward t an object in the Kuiper Belt covered by living creatures, the brightness of the object nay vary noticeably as the Earth moves nearer or farther from the axis of the reflected eam. It will be brightest when the Earth is directly between the sun and the object.

Home-based pit-lamping is not a substitute for orbital pit-lamping. The two kinds f pit-lamping complement each other without much overlapping. Both are needed if e are to extend our search for life as widely as possible.

Orbital pit-lamping can search for small and sparse patches of life on a few objects uch as Europa. Home-based pit-lamping can search for big and luxuriant patches of fe covering the surface of a big number of more distant objects. Orbital pit-lamping n Europa is like searching for lichens in rocks in Antarctica with a hand-held nagnifying glass. Home-based pit-lamping in the Kuiper Belt is like searching for rain-orests on Earth with a camera on the Moon.

To discuss the possibilities of home-base pit-lamping in a quantitative way, I need) introduce some units. Since we are talking about astronomical observations, it is onvenient to use the astronomical unit of apparent brightness which is the magnitude. he magnitude is a logarithmic measure of brightness. Bigger magnitude means smaller ntensity of light. For those of you who are not amateur astronomers, here are the nagnitudes of the planets in the outer part of the solar system: Jupiter minus 2, Saturn lus 1, Uranus 6, Neptune 8, Pluto 15. Adding one to the magnitude means making the ight fainter by a factor of two and a half.

Now suppose that Pluto itself were covered with living creatures with optical oncentrators. Then its light would be brightened by a factor of 2,500 which means ight and a half magnitudes. Pluto would be magnitude six and a half, almost as bright s Uranus, and brighter than any of the asteroids except for Vesta. Pluto would have een discovered two hundred years ago when the first asteroids were found. So we can e sure that Pluto is not covered with exotic sunflowers gazing at the distant sun.

Pluto is a unique object, the biggest so far discovered in the Kuiper Belt, with liameter 3,000 kilometers and magnitude 15. The faintest objects we can identify in the Kuiper Belt with ground-based telescopes are about a thousand times fainter, with nagnitude 22.5 and diameter about 100 kilometers. With the Hubble telescope we can letect objects about 5 magnitudes fainter, to magnitude 27.5 and diameter 10 kilometers, ut we don't usually get enough observing time on Hubble to determine the orbit of an ndividual object and identify it as belonging to the Kuiper Belt. The practical limit for

identifying Kuiper Belt objects is magnitude 22.5, until we have a space telescope wit
large amounts of time dedicated to this purpose.

Suppose one of the smaller objects at the inner edge of the Kuiper Belt is covere
with sunflowers. Suppose that it is barely detectable, with magnitude around 22.5. Tha
means that without the sunflowers it would have had magnitude 31, 16 magnitude
fainter than Pluto, far too faint to be detected even by Hubble. With the sunflowers, i
it has the same albedo as Pluto, it would be 1,500 times smaller than Pluto. Its diamete
would be about 2 kilometers. The home-based pit-lamping strategy enables us to detec
any object at the inner edge of the Kuiper Belt that has as much as 3 square kilometer
of area covered with sunflowers. And this is not all. The Kuiper Belt extends far beyon
the distance of Pluto, and beyond the Kuiper Belt there is the Oort Cloud, containin
billions of objects orbiting the sun at distances extending out further than a tenth of
light year. Pit-lamping becomes more and more effective, the further out you go.

For non-living objects shining by reflected sunlight, the brightness varies with th
inverse fourth power of distance, two powers of distance for the sunlight going out an
another two powers for the reflected light coming back. For living objects with optica
concentrators, the concentration-factor increases with the square of distance t
compensate for the decrease in sunlight. Then the angle of the reflected beam varie
inversely with distance, and the intensity of the reflected beam varies with the invers
square instead of the inverse fourth power of distance. Because the linear size of th
reflected beam is independent of the distance of the object, the Earth will remain withi
the beam no matter how far away the object sits. If an object is at a tenth of a light yea
distance, the concentration-factor will be 100 million, and the angle of the reflecte
beam will be about 20 seconds of arc. This requires optical quality a little better tha
the human eye but much less precise than an ordinary off-the-shelf amateur telescope
At that distance, we could identify an Oort Cloud object covered with sunflowers if it
diameter were as large as 400 kilometers. This diameter is much smaller than Pluto an
comparable with other Kuiper Belt objects that were discovered in the last few years.

The prospects for pit-lamping look even more promising if we look at the situatio
not from our Earth-bound point of view but from the point of view of the life that we ar
searching for. A community living on the surface of a small object far from the sun ha
two tools available to increase its chances of survival. One tool is to grow optica
concentrators to focus sunlight. The other tool is to spread out into the space around th
object, to increase the area of sunlight that can be collected. So far, we have onl
considered the effect of optical concentrators. We next consider the effect of increasin
the area by spreading out leaves and branches into space.

Imagine an ecological community containing a large variety of species ranging from bacteria to plants and animals, comparable with a rain-forest on Earth, and occupying one square kilometer of area. It requires for its sustenance an input of 200 watts per square meter of sunlight, or 200 megawatts for one square kilometer. Now imagine this community installed on a little comet of diameter one kilometer and moved to the Kuiper Belt. It must grow optical concentrators, as we have seen, to concentrate sunlight by a factor of 2,500 and keep its absorbing surfaces at a comfortable temperature. But it must also increase its absorbing area by a factor of 2,500 to keep its total supply of energy constant. It must grow out into space to form a disc of diameter 50 kilometers instead of one kilometer. The gravity of an object of this size is so weak that it imposes no limit on the distance to which a life form such as a tree can grow. The living community with diameter 50 kilometers, with its optical concentrators spread over the enlarged area, will then be visible from Earth as a speck of light of magnitude 15, about as bright as Pluto.

Now suppose that the same living community is moved out to the Oort Cloud at a distance of a tenth of a light year, a factor 200 further away than Pluto. To maintain the same life-style with the same population of creatures, it must concentrate sunlight by a factor of 100 million and also increase its area to 100 million square kilometers. It will grow into a thin disc of diameter 10,000 kilometers, covered with big flimsy mirrors to concentrate the light. The original comet with diameter 1 kilometer will have a mass around one billion tons. To make an optical concentrator, using thin film reflectors one micron thick, requires about one gram per square meter or one ton per square kilometer. With an area of 100 million square kilometers, the community needs to use only one tenth of the mass of the comet to make the reflectors, with the other nine-tenths left over to use for its other activities and for building structures. Now comes the remarkable conclusion of this calculation. The community in the outer region of the Oort Cloud is still visible to us as a magnitude 15 object. The inverse fourth power dependence of brightness on distance, which holds for non-living objects, can be completely canceled for living objects. Living communities can cancel two powers of distance by their use of optical concentrators, and can cancel two more powers of distance by their increase in area. If the rate of consumption of energy by a community is fixed, the brightness of the community as seen from the Earth is independent of distance. We can easily detect any community that uses a hundred megawatts of sunlight or more for its life-support, out to a distance of a light year from the sun. Beyond a light year, the sun is outshone by Sirius. Anything living beyond a light year from the sun would point its concentrators at Sirius rather than at the sun. At that point, home-based pit-lamping no longer works, unless we are thinking of alien pit-lamping astronomers in orbit around Sirius, with Sirius as their home base.

The practical upshot of these calculations is that it would make sense to search for life on the surface of objects in the outer solar system with ground-based telescopes. All we need to do is look for objects in the Kuiper Belt or the Oort Cloud with unreasonable brightness. Anything with parallax and proper motion appropriate to the Kuiper Belt or the Oort Cloud, and with brightness greater than expected for an object at that distance, would be a candidate for an inhabited world. If we find such objects and they do not turn out to be inhabited, they would still be important discoveries. They would be objects of outstanding size, a new family of giant comets or small planets, in a remote part of the solar system. If such objects exist, either inhabited or uninhabited, it may already be possible to find them by looking at the output of the Sloan Digital Sky Survey or other sky surveys. Any interesting candidates could be examined more carefully with larger telescopes.

3. Life in Vacuum

I want now to talk in a more general way about the possible history and evolution of life if it happens to exist away from planets. We have seen on planet Earth that life has two outstanding qualities that distinguish living from non-living objects. These qualities are adaptability and invasiveness. During the four billion years that life has existed on Earth, it has adapted itself to an amazing variety of ecological niches with an amazing variety of ways to make a living. It has invaded the most inhospitable and inaccessible corners of the planet, from frozen rocks in Antarctica to thermal vents in the depths of the oceans. It is reasonable to expect that life, if it exists elsewhere, will still be characterized by extreme adaptability and invasiveness. In an unstable and unpredictable universe, these qualities are needed if life is to survive, either on a planet or anywhere else.

The most important invasion in the history of life on Earth was the move from the ocean onto the land. This invasion was not easy. Life was confined to the oceans for nearly three billion years before it learned how to survive on land. We do not know precisely how or when the invasion began. It probably began with microbes, followed later by plants and then by animals. The fact that the Earth has an atmosphere was helpful to the invaders in at least five ways. The atmosphere protected the invaders from lethal ultraviolet radiation from the sun. It also provided shielding from cosmic rays. For the plants, it provided carbon dioxide and molecular nitrogen for their metabolism. For the animals it provided molecular oxygen for the more rapid metabolism demanded by brains and muscles. And finally, it carried clouds and rain which provided life-saving water to replace the ocean that the invaders had left behind. The invaders made full use of these resources provided by the atmosphere as they slowly adapted to the new environment.

Now I am speculating that if life had originated on an airless satellite such as Europa instead of on a planet with air, the move from ocean to vacuum would not necessarily have been more difficult than the move from ocean to air. Let us consider in turn the five resources provided by an atmosphere. The shielding from ultraviolet radiation would hardly be needed on Europa since sunlight is very weak, and if shielding were needed it could be provided by a thin layer of opaque skin or bark. The shielding from cosmic rays would not be needed if the invaders evolved the same resistance to radiation damage that many species of microbes and insects evolved on Earth. The carbon, nitrogen and other elements required for metabolism of plants are probably available in the dark stripes that we see on the surface of Europa, where brine from the ocean underneath seems to have emerged from cracks in the ice. Water is certainly available in unlimited quantities anywhere on the surface. Finally, the oxygen required by animals poses the most difficult problem, but there are many ways in which this problem may be side-stepped. One possible way was imagined by Konstantin Tsiolkovsky more than a hundred years ago in his book, *Dreams of Earth and Sky*, [Tsiolkovsky, 1895].

Tsiolkovsky's vacuum-dwellers were not divided into plants and animals, but combined the characteristics of both. He called them animal-plants. Here one of them is explaining how his metabolism works: "You see these green appendages on our bodies, looking like beautiful emerald wings? They are full of chloroplasts like the ones that make your plants green. A few of your animals have them too. Our wings have a glassy skin that is airtight and watertight but still lets the sunlight through. The sunlight dissociates carbon dioxide that is dissolved in the blood that flows through our wings, and catalyzes a thousand other chemical reactions that supply us with all the substances we need." If you are an animal-plant, the oxygen released by photosynthesis in your wings is carried in your blood to your brain and muscles, so you do not need air to supply it. If you were living on the surface of Europa, you would not have needed to invent lungs. Tsiolkovsky imagined many other clever tricks by which life might have adapted to the perils and opportunities of the vacuum environment. Lacking air to transmit sound, his animal-plants found a better way to talk to each other. "One part of their body carries under the transparent skin an area like a camera obscura, on which moving pictures are continually playing, communicating by following the flow of their thoughts and representing them precisely. The pictures are formed by fluids of various colors which flow through a web of fine channels under the skin."

Let us suppose that life has emerged onto the surface of some body without an atmosphere and has developed a robust ecology of creatures adapted to living in vacuum. This might have happened on Europa or on Callisto or anywhere else in the outer solar system. Now comes the most important advantage of vacuum life as compared with air-breathing life. It is far easier for vacuum life to spread from one world to another than it is for air-breathing life.

The reason is simple. If a chunk of ice is knocked off Europa by a cometary impact, and if the chunk has vacuum-life on its surface, then the life has a good chance of surviving the impact and continuing to flourish while the chunk wanders around the solar system. The environment after the impact is not much different from the environment before. And if the chunk happens to land on another object with not too high a relative velocity, the life has a chance of surviving the second impact and establishing itself on another world. This transition from one world to another is far more difficult for air breathing life. Joe Kirschvink, [Kirschvink, 2000], has shown that it is possible that life could have spread from Mars to Earth in the interior of a chunk of rock knocked off Mars and landing on Earth. He studied the patchy magnetization in thin slices of the famous Mars rock ALH84001 which was picked up in Antarctica, and demonstrated that the magnetization would have disappeared if the interior of the rock had ever been heated above 40 degrees Celsius during its exit from Mars or during its arrival on Earth. If a living bacterium or spore had been hiding inside that piece of rock, it could conceivably have survived the trip and emerged to populate the Earth with its descendants. But any living passengers in rocks traveling from Mars to Earth must have been microbes or spores in a state of suspended animation. Air breathing creatures could not have come to Earth in this way. Only if life is adapted to vacuum could a whole ecology of more advanced creatures perhaps make the trip together from one world to another.

Europa, as it happens, is one of the more difficult places for life to escape from, because of the large gravitational potential of Jupiter. The escape velocity from Europa itself is less than 2 kilometers per second, but the escape velocity from Jupiter starting from the orbit of Europa is about 6 kilometers per second. If a chunk of ice escapes Europa but does not escape Jupiter, it will continue to orbit Jupiter in a Europa-crossing orbit, and it is likely to crash into Europa again before it encounters any other possible destination. To escape from Jupiter with a velocity of 6 kilometers per second departing from Europa, it must have the good luck to leave in the right direction so that its velocity is added to the orbital velocity of Europa. If surface life exists on Callisto, escape from Jupiter would be easier but still not probable. Callisto itself has an escape velocity of 2.3 kilometers per second, a little larger than Europa, and the escape velocity from Jupiter starting from Callisto's orbit is 3 kilometers per second instead of 6. It would still need an improbable combination of circumstances for life on Callisto to escape from Jupiter. But hopping from world to world becomes rapidly easier as we move out beyond Jupiter to the Kuiper Belt and the Oort Cloud.

If surface life exists on a typical object in the Kuiper Belt with a diameter of a few kilometers, the escape velocity will only be a few meters per second. Orbital motions in the Belt are slow, and relative velocities of neighboring objects are typically around 1 kilometer per second or less. Gentle collisions between neighboring objects will be common. Life has a good chance of surviving when a gentle collision knocks a piece of its home into space, and when the piece later makes a gentle collision landing on another object. In this way, life could spread from world to world like neutrons in a divergent nuclear chain reaction. A large fraction of the objects in the Kuiper Belt might become inhabited. If this should happen, the process of Darwinian evolution would then select life forms that are particularly well adapted to traveling. Life forms might evolve that do not require a chance collision to leave their home but spontaneously hop into space when conditions at home become crowded, taking with them enough material resources to survive independently. Such life forms would be winners in the race to colonize new worlds.

Life forms that do not hop, but grow far out into space around their home territory, would also have an advantage in spreading to other worlds. In addition to acquiring more sunlight, they would acquire a bigger cross-section for collisions with other Kuiper Belt objects. When life has grown out into a thin disc with an area thousands of times larger than its original territory, the effect of a collision will usually be to punch a small hole in the disc, with minor damage to the life that stays behind, and with a substantial chance of transferring seeds of life to the object that punched the hole. In the Kuiper Belt there is probably a substantial quantity of matter in the form of dust and ice crystals, too small and too dispersed for our telescopes to detect. By growing out into space, life forms could greatly increase the amount of this material that they encounter, and could collect and use it for their own metabolism. Life forms that adopt this strategy would be like the filter-feeders that we find in tide pools on Earth. Other life forms in the Kuiper Belt might adopt a more active strategy, using eyes to locate larger objects floating by with small relative velocity, then hopping into space to intercept them. When life has colonized a substantial fraction of Kuiper Belt objects and the competition for real estate becomes intense, we may expect to see life forms diversifying into predators and prey, carnivores and herbivores. In the wide spaces of the Kuiper Belt, evolution will drive life to take maximum advantage of the occasional collisions and catastrophes that punctuate its otherwise quiet existence. Life will be driven by Darwinian competition toward maximum adaptability and invasiveness, the same qualities that competition nurtured on our own planet.

Beyond the Kuiper Belt lies the Oort Cloud, where the distances between habitable objects are larger and the relative velocities smaller. In the Oort Cloud, relative velocities will be of the order of a hundred meters rather than a kilometer per second. Here it will be even easier for life to hop from one object to another. We know very little about the total number of these objects, but we know quite a lot about their chemical and physical constitution, since these are the objects that we see as long-period comets when they are occasionally deflected by gravitational perturbations into orbits that pass close to the sun. Then we see them boil off tails of gas and dust that we can observe and analyze. We know that they typically have diameters of a few kilometers and are largely composed of the biologically essential elements, hydrogen, carbon, nitrogen, oxygen. Since the reservoir of these objects in the Oort Cloud delivers about one object per year into the inner Solar system where we can observe it as a comet, and since the system is four billion years old, the reservoir must contain at least several billion objects of kilometer size.

It is not likely that the Oort Cloud, a vast desert of space with habitable oases separated by huge distances, is actually inhabited. But it might conceivably be inhabited by life forms with sufficiently precise mirrors to concentrate sunlight by a factor of a hundred million. The remarkable fact is that the strategy of home-based pit-lamping could enable us to detect such life forms if they exist.

4. Beyond the Solar System

Next I will talk briefly about the possibilities of vacuum life existing beyond the solar system. There is a radio astronomer named Jack Baggaley at the University of Canterbury in New Zealand who uses a radar system called AMOR (Advanced Meteor Orbit Radar) to track incoming micrometeors that leave trails of ionized plasma in the upper atmosphere, [Baggaley, 2000]. The tracking is accurate enough so that he can determine the velocities of individual objects with a three-sigma error of 10 kilometers per second. The objects that he can detect have diameters around 20 microns, much larger than normal interplanetary dust particles, much smaller than normal meteorites that fall to the ground. He finds that a substantial fraction of these objects have hyperbolic velocities relative to the sun, so that they do not belong to the solar system but are coming in from outside. He finds a second fact, which is not so firmly established, but still appears to be true. Of the objects that come from outside the solar system, a substantial fraction are coming from a single direction, roughly from the direction of the star Beta Pictoris, known familiarly to astronomers as Beta Pic.

Beta Pic is famous because it is a bright star, only 63 light years distant from the sun, with a very large disc of dust orbiting around it. The disc around Beta Pic is estimated to contain about a thousand times as much material as our Kuiper Belt. The dust grains that Baggaley observes coming from Beta Pic are our first direct evidence that objects are moving from one system to another. The dust grains are presumably thrown out of the Beta Pic system by gravitational encounters with Beta Pic planets that we have not yet observed. And if this happens to dust grains, it presumably also happens to kilometer-sized objects. We must expect that among the billions of objects in our Oort Cloud there will be a substantial population of kilometer-sized interlopers passing through on their way from Beta Pic. No such interloper has been seen among the long-period comets that we observe passing close to the sun. All the comets whose orbits have been accurately measured belong to our solar system. But the total number of observed comets is small, and the absence of observed interlopers is consistent with the presence of millions of unseen interlopers passing through the Oort Cloud.

Another radio astronomer, John Mathews at Pennsylvania State University, observing with the radar telescope at Arecibo in Puerto Rico, has also found dust grains arriving from outside the solar system, but he disagrees with Baggaley about the direction from which they are coming, [Meisel, Janches and Mathews, 2002]. Mathews finds a substantial fraction coming from the direction of a famous object called Geminga, a supernova remnant which is also a source of gamma-rays in the constellation Gemini. Geminga is an enigmatic object with no resemblance to Beta Pic. I will not try to resolve the dispute between Mathews and Baggaley. It seems likely that they are both observing the same stream of dust grains. Beta Pic and Geminga are at the same ecliptic longitude but differ in ecliptic latitude. Beta Pic is close to the latitude of New Zealand, Geminga is close to the latitude of Puerto Rico. It is easy to imagine that observational bias resulted in the different identifications of the source of the dust grains. To avoid further argument, I shall arbitrarily assume that at least some of the grains are coming from Beta Pic.

If life can exist in our Kuiper Belt, then much more life might be flourishing in the bigger and denser disc around Beta Pic. And if kilometer-sized objects are leaving the Beta Pic system and arriving in our solar system, it is possible that Beta Pic life is passing through our Oort Cloud as we speak. It is then an interesting question, whether we could detect life on Beta Pic objects in the Oort Cloud using home-based pit-lamping. We assume that the Beta Pic creatures direct their optical concentrators toward the sun as soon as they come within range. The problem is that the Beta Pic objects, unlike the Oort Cloud objects, are likely to be moving at high velocities transverse to our line of sight. Sunlight that impinges on a Beta Pic object will not be reflected straight back at the sun. The reflected light will be deflected by the angle $(2V/c)$, where V is the transverse velocity of the object and c is the velocity of light. The conical reflected beam has half-angle (a/D), where a is the Earth-Sun distance and D is the distance to the object. According to Baggaley's measurements, a typical transverse velocity is 50 kilometers per second, which gives a deflection angle of $(1/3,000)$ radian or one minute of arc. The reflected beam will miss the sun if the distance D is greater than 3,000 times a, or greater than a twentieth of a light year. This leaves a very substantial volume within which detection of life on Beta Pic objects is in principle possible.

If we are to take seriously the possibility of life traveling from one solar system to another, two more questions need to be addressed. First, is it possible for life to survive the transit using ambient starlight as it moves through the galaxy? To answer this question, we remark that here on Earth we have three stars brighter than zero magnitude in the sky, not counting the sun, namely Sirius, Canopus and Alpha Centauri, and a couple of others, Arcturus and Vega, that are within a tenth of a magnitude of zero. If our present situation in this part of the galaxy is not exceptional, we can safely assume that on the way from Beta Pic to here we always can see at least one star in the sky that is zero magnitude or brighter. If the life in transit uses optical concentrators weighing one gram per square meter of area, then the light from a single zero magnitude star provides 0.02 watts of available energy per ton of concentrators. If the object in transit is an average comet weighing a billion tons and uses half its mass for concentrators, then the available energy is 10 megawatts. This is enough energy to sustain a modest community of living creatures as they cruise across the galaxy. It would be enough to sustain a village of a few hundred human beings with a modern western standard of living.

The second question that needs to be addressed is whether, after an inhabited object arrives from Beta Pic with high velocity, it is possible for it to become a slow moving object belonging to our own solar system? The answer to this question is again affirmative. Although the great majority of fast moving objects will pass through the Oort Cloud without any interaction, a few will be deflected by close encounters with planets. A fraction of these few will be captured into orbits around the sun, and a smaller fraction will be deflected by further gravitational encounters into orbits indistinguishable from orbits belonging to our own Kuiper Belt or Oort Cloud. From that point onward, the alien life would be at home in the solar system and could spread to other Kuiper Belt or Oort Cloud objects as if it were native. Having answered both these questions in the affirmative, we can assert that life adapted to vacuum has the potential to spread from its place of origin, not only from world to world within our solar system, but far and wide through the galaxy. Life adapted to an atmosphere is stuck on the planet where it started.

If we look at the universe objectively as a home for life, without the usual bias arising from the fact that we happen to be planet dwellers, we must conclude that planets compare unfavorably with other places as habitats. Planets have many disadvantages. For any form of life adapted to living in an atmosphere, they are very difficult to escape from. For any form of life adapted to living in vacuum they are death-traps, like open wells full of water for a human child. And they have a more fundamental defect: their mass is almost entirely inaccessible to creatures living on their surface. Only a tiny fraction of the mass of a planet can be useful to its inhabitants. I like to use a figure of demerit for habitats, namely the ratio R of total mass to the supply of available energy. The bigger R is, the poorer the habitat. If we calculate R for the Earth, using total incident sunlight as the available energy, the result is 12,000 tons per watt. If we calculate R for a cometary object with optical concentrators, traveling anywhere in the galaxy where a zero magnitude star is visible, the result is, as we have seen, 100 tons per watt. The cometary object, almost anywhere in the galaxy, is 120 times better than planet Earth as a home for life. The basic problem with planets is that they have too little area and too much mass. Life needs area, not only to collect incident energy but also to dispose of waste heat. In the long run, life will spread to the places where mass can be used most efficiently, far away from planets, to comet clouds or to dust clouds not too far from a friendly star. If the friendly star happens to be our sun, we have a chance to detect any wandering life form that may have settled here.

5. Political Coda

This chapter has been about science and technology. Let me end it with a few remarks about the politics of space missions. I do not use the word politics in a derogatory sense. So long as space missions are paid for by tax payers and funded by politicians who represent the tax payers, politics are an essential and healthy part of the business of exploring space. Science and technology provide the means for exploring space. Politics decide which goals are worth the cost of exploring. As a general rule, a space program flourishes when there is political agreement about goals, and fails to flourish when there is political disagreement. At the present time, there are signs of a dangerous division of opinion about the goals of exploring space beyond the Earth-Moon system. On the one side, there are the astronomers and planetary scientists whose goal is to understand the physical and chemical processes going on in the universe as widely as possible. On the other side, there are the professional astrobiologists and the amateur enthusiasts whose goal is to search for evidence of extraterrestrial life. To keep the program healthy, it is essential to fly missions that satisfy both these goals, so that they will generate political support from both sides. It will be much easier to find such dual-purpose missions, if the search for life is not narrowly construed to mean searching on Mars and Europa only. The wider our search for life, the easier it will be to combine it with other exploratory objectives. In addition to the scientific reasons for expecting life to be detectable away from planets, there is also a strong political reason to search for it in unlikely places.

What is the moral of this story for our space program? The chief moral that I draw is that all-sky surveys are a good idea. We need to fly more all-sky missions like IRAS, COMPTON, MAP, not to mention ground-based surveys like 2MASS and Sloan Digital Sky Survey. We may hope to find life with searches targeted on planets and their satellites, on Mars and Europa in particular, but we may have a better chance to find evidence for life in unexpected places if we search in all directions with an all-sky survey. I will not be surprised if the first extraterrestrial life to be discovered turns up in a part of the sky where there is nothing to be seen but a couple of dim Kuiper Belt objects with anomalously small parallaxes and anomalously slow proper motions. I will also not be surprised if the first sign of life is an anomalous infrared emission from something that looks like a small interstellar dust cloud. The main practical lesson that I want to leave with you is this: Keep on looking for life in unexpected places, and especially in places where we may find clues to other scientific mysteries that have nothing to do with life.

References

o Baggaley, W. Jack, 2000. "Advanced Meteor Orbit Radar observations of interstellar meteoroids," J. Geophys. Research,105, 10353-10361.

o Kirschvink, Joseph L. *et al.*, 2000. "A low temperature transfer of ALH84001 from Mars to Earth," Science, 290, 791-795.

o Meisel, David D., D. Janches and J.D. Mathews, 2002. "Extrasolar micrometeors radiating from the vicinity of the local interstellar bubble," Astrophys. J., 567, 323-341.

o Tsiolkovsky, Konstantin, 1895. "Dreams of Earth and Sky," [Moscow, Goncharov]. Edition edited by B.N. Vorobyeva, [Moscow, USSR Academy of Sciences, 1959], pp. 40-41.

† Talk given at Jet Propulsion Laboratory, May 6, 2002

— PART IV —

Memories of Charles Sheffield and Robert Forward

Remembering Charles Sheffield

Yoji Kondo / Eric Kotani

The first time I heard the name Charles Sheffield was at a dinner with Robert A. Heinlein in July 1979. Heinlein was testifying at the joint session of the Congress and had invited me to dine with him, his wife, Ginny, and his agent, Eleanor Wood. Over the dinner, he asked what science fiction books I had been reading lately. I answered, somewhat reproachfully, that I had been reading little since he had not been writing much. He said he must fix my diet, and gave me the names and books of several authors that he thought well of. Among them was *Sight of Proteus* by Charles Sheffield. I met Charles later that year at the party hosted by Heinlein in Annapolis at one of his alumnus meetings. Jim Baen, who too was at the party, talked us into writing an article for his magazine, *New Destinies*. How he trapped us into writing *Looking about in Space* is another story, but I am certainly glad that he did. Within a few years, I was writing science fiction novels with John Maddox Roberts. Charles and I occasionally had lunch together at a local Chinese restaurant, and we visited each other's home from time to time.

Charles had a Ph.D. in mathematical physics. His undergraduate mentor at Cambridge was Fred Hoyle; Charles once deciphered Hoyle's handwritings for me for an astrophysical magazine I edited. When he was not writing science fiction, he was Chief Scientist for Earth Satellite Corporation for a number of years; we shared many common interests in the exploration of space. He helped me co-organize two AAAS (American Association for the Advancement of Science) symposia, *Space Access and Utilization beyond 2000* (Washington, D.C., in 2000), and *Interstellar Travel and Multi-Generation Space Ships* (Boston in 2002). A book based on the first meeting was published from the American Astronautical Society with Charles as a co-editor, and a new book based on the second conference, for which Charles has written an article, you hold in your hands.

The last time I spoke with him was a few months ago, just one day before he went in for his operation. We discussed the Heinlein Award, which he had helped me set up, and then he told me about his medical problems. True to his form, he sounded as usual. Little did I imagine that it was to be my last conversation with him.

Charles Sheffield was a fine writer, a good scientist, and a wise and trustworthy colleague. He was always ready to render help to his friends and to new writers as well. My life has been enriched by knowing him as a friend.

Bob Forward: Reminiscences

Geoffrey A. Landis

I first saw Bob Forward in 1980, in Boston; although, of course, he wouldn't know that.

I wasn't actually attending the world science fiction convention; I wouldn't attend my first SF convention for another five years. That year I was working hard on building models and getting some flying time in, preparing for the space modeling world championships in another two weeks, and even though the science fiction convention was nearby (walking distance from the MIT Rocket Society's headquarters on the Charles River) I was far too busy to think about actually getting a membership. But I was curious about what this "science fiction convention" thing was about, so I crossed the river one afternoon and walked in to a panel at random.

On the panel I walked into – "Worldbuilding" – Bob Forward was talking with a wicked grin about the process he went through in writing *Rocheworld*, and telling how he had to invent a new way to do interstellar flight, since he'd casually given his earlier idea away to Larry Niven and Jerry Pournelle two years before, and they'd written it into a novel. The way he spoke, it sounded a lot more like fun than work.

The interstellar flight system he was talking about eventually appeared in a novel (which I read in the *Analog* serial under the title *Flight of the Dragonfly*), and then a few years later showed up as a technical paper with some of the details worked out. The paper, as it turned out, was a real landmark in the field; it was the first in-depth technical discussion of a method of interstellar flight that didn't rely on hand-waving physics, and with detailed numbers, he showed that it had a realistic chance of working. By this time I had quit my job for graduate school, and the month Bob's paper came out in *Journal of Spacecraft and Rockets*, I had just submitted my first story to *Analog*, emboldened by seeing from Bob's example that it was indeed possible to be both a physicist and also a science fiction writer.

I studied the paper and made some of my own calculations (one of my friends at the time told me that if I'd spent as much time working on my research as I spent doing research on interstellar flight, I would have graduated a year earlier). A little later I did graduate, though, and started on a postdoc at NASA Lewis Research Center (now renamed NASA Glenn). I discovered that NASA would send me to technical conferences if I was presenting a paper, and discovered that the International Astronautical Federation Congress actually had a session devoted to interstellar flight.

By this time I had a lot of further thoughts about Bob's laser-pushed light sail, and I thought that it might be worthwhile to put them together into an actual paper, which I did (*Optics and Materials Considerations for a Laser Pushed Light Sail*, paper IAA-89-664.) I had noticed what seemed, at the time, to be some difficulties with his original concept, and the paper discussed these and suggested some alternate ideas, including the concept of making the sails out of a thin dielectric film, such as diamond, or of using a particle beam instead of a laser beam to push the sail. I sent a copy to Bob, and got back a long and detailed discussion, neatly written in purple felt marker. Some of my comments he praised, but on a couple of other points, where I'd thought he'd made an error, he took issue and eventually explained his thinking in simple enough detail for me to follow it, and after a while I was forced to admit that I had been in error: coherence does, in fact, allow a laser beam to focus to a distant spot size that is smaller than the optically-magnified image of the exit aperture of the laser.

I hadn't realized it at the time, but with the paper I had joined a rather small and exclusive fraternity: the gang of people who were seriously thinking about how to do technically realistic interstellar flight – a gang in which Bob Forward was, beyond any challenge, the acknowledged ringleader.

About that time Marc Millis and I and a few others at NASA Lewis were working on ways for getting NASA to be more accepting of advanced thinking. Marc had started up a group at Lewis called Vision-21, short for "visionary thinking for the twenty-first century," and I was one of the charter members of the club. One of the things we wanted to do was to have a symposium for advanced thinking at Glenn, inviting a lot of the people who were comfortable with wild ideas to get together, and see whether we could make some sparks. We held the first symposium, *Vision 21: Space Travel for the Next Millennium*, in April 1990, and the top of our list of people to call to give an invited talk was Bob Forward, as a premier example of exactly the kind of advanced-concepts thinking that we wanted NASA to do more of. When he gave his talk, he started out by saying that his original paper was going to review way-out ideas for space flight, but he got to thinking about it, and decided that there were already enough far-out ideas in the literature, and since the symposium was being held at NASA, it would be more useful to instead point out ideas that NASA could implement in the near term. The topic that he picked to talk about in detail was the idea of using tethers to transfer momentum, something that, he said, could be implemented in the near future, without waiting for breakthroughs or new materials or stretching the boundaries of physics, but yet might demonstrate ideas that could eventually revolutionize space flight. (I'd been corresponding with him for a while on a variety of subjects, and I was pleased to see that he discussed some of my ideas on the paper he eventually submitted to the proceedings.) One of the points that he discussed was the fact that it would be necessary to design a tether to avoid catastrophic failure in the event of an impact by space debris; he said that this ought to be doable, but it would need study. He then went on to show an example of how a tether system could be used for a rocketless Earth-to-Moon transportation system.

Of course, those two points – tether transportation systems that could be implemented in the near term, and the development of tether systems that can sustain multiple cuts without failure – went on to be a major thread of the most recent decade in Bob Forward's remarkable career. I like to think that our little symposium gave at least a little bit of "forward momentum" to that development.

By then my orbit and Bob's went on to intersect at many points, meeting at advanced-propulsion workshops and beamed-energy symposia. Usually I discovered that I was trailing about twenty degrees behind, and half of the ideas that I came up with had been mentioned as a throw-away line in a paper by Robert Forward bursting with other ideas. Watching the way that Bob developed ideas, and seeing how he could ask the right questions and pull the important new concept out of a mass of information, has been a continuing learning experience in how to get new ideas taken seriously.

When Bob was unable to come to Boston for the AAAS meeting, and asked me to present his paper for him, I have to admit to being extremely flattered. After all, it's a tough job to be asked, even temporarily, to step into the shoes of Bob Forward!

I have to salute Bob Forward. In the long term, we're headed for the stars, and Bob Forward was the pioneer, step by step charting the uncharted territory and looking for the science behind the science fiction.